An Engineer's Introduction to Programming with MATLAB 2017

Shawna Lockhart

Eric Tilleson

Publications

SDC Publications
P.O. Box 1334
Mission, KS 66222
913-262-2664
www.SDCpublications.com
Publisher: Stephen Schroff

ISBN-13: 978-1-63057-125-2

ISBN-10: 1-63057-125-3

Printed and bound in the United States of America.

PREFACE

Welcome to An Engineer's Introduction to Programming with MATLAB®! We truly hope that this book is a useful resource in introducing you to the extraordinarily powerful MATLAB® software and in teaching you basic computer programming that will serve you regardless of your programming environment.

We've written this book thinking of students who have little or no programming experience, and possibly little computer experience. Some of the early concepts might be second nature to those of you with experience, but by the end you'll be well into new territory.

As you work your way through the tutorials, we can't encourage you enough to experiment beyond our instructions. Click buttons that look interesting, type in variations of commands, type commands incorrectly on purpose just to see what happens, examine the menus, read the MATLAB help for extra options; you'll learn so much from exploring on your own. Don't worry, you won't permanently break anything, and most people agree that a person learns more from making a mistake than from getting it right the first time.

Key Features

This book includes:

- Step-by-step tutorials written for the novice user to become proficient using MATLAB

- Getting Started chapter for configuring MATLAB for use with the tutorials

- Organization and level suitable for a first year introductory engineering course

- Tips offering suggestions and warnings as you progress through the book

- Key Terms and Key Commands listed to recap important topics and commands learned in each tutorial

- An Index to help you look up topics

- Exercises at the end of each tutorial providing challenges to a range of abilities.

Exercises

The exercises included at the end of each chapter give you the chance to practice using MATLAB. It is essential to try out the information in the chapter by working through the exercises. You will learn to find your own programming style and your approach to solving problems using MATLAB. In addition you will gain basic troubleshooting skills.

Acknowledgments

The authors would like to thank Anthony Demetriades, phD for sharing his years of experiences in university teaching and research to keep us on the straight and narrow, suggesting chapter exercises, and reviewing the text.

Thanks also to Gabe Taurman for being our stand-in for students one and all, working his way through the book and providing invaluable feedback. Lastly, our thanks to Stephen Schroff, Karla Werner, and Zach Werner at SDC Publications for their support and for getting our book into your hands!

Happy learning!

Eric Tilleson

Shawna Lockhart

ii

BRIEF TABLE OF CONTENTS

TABLE OF CONTENTS

GETTING STARTED

Preparing MATLAB® for the Tutorials

The tutorials in this guide will teach you to use the MATLAB® 2017 software for Windows operating systems through a series of step-by-step exercises.

We assume you have successfully installed the software with its default settings and that you have activated the software.

In this book, we show MATLAB running under Windows 10 operating system. Windows 7 and Windows 8 installations will look very similar to that of Windows 10. MATLAB takes advantage of common Windows file operations and allows you to run the MATLAB software while running other applications.

If you are using a different operating system, the tablet version, or on-line version of MATLAB, the instructions in the book will not always match your user experience exactly. That said, you will still probably be able to figure it out.

Basic Mouse Techniques

We assume you are using a 2-button mouse with a middle roller-button. The following terms describe use of the mouse. (These directions are given for right-handed mouse buttons.). You can also use a touch screen in a similar fashion to the instructions given for typing from a keyboard and using a mouse.

Term	Meaning
Pick or Click	Quickly press and then release the left mouse button
Right-click	Quickly press and release the right mouse button
Double-click	Rapidly click the left mouse button twice
Drag	Press and hold down the left mouse button while you move the mouse
Highlight	Move the mouse until the mouse pointer/cursor is positioned to indicate the item you want
Select	Position the mouse pointer/cursor over an item and click the left mouse button.

Recognizing Typographical Conventions

During these tutorials, you'll use your keyboard and mouse to input information. As you read, you'll see special type styles to help you identify the information to input. Some of the special type illustrates computer keys, for example, the Enter key is represented by [Enter].

Some special typefaces present instructions you will perform. These instructions are indented from the main text and indicate a series of

Objectives

This chapter describes how to prepare your computer system and the MATLAB software for use with the tutorials in this manual. As you read this chapter, you will

1. Understand how to use a mouse during the completion of these tutorials.

2. Recognize the typo-graphical conventions used in this book.

3. Create and prepare data file and working directories.

4. Configure your copy of MATLAB for use with these tutorials.

Tip: *You can use the Mouse Properties selection in the Windows control panel to switch your mouse to a left-handed configuration. This will switch your pick button to the right button on the mouse. The left mouse button will then act as the enter or return button.*

Note: *The screens in this book were captured on a touchscreen Microsoft Surface Pro 4 computer running Microsoft Windows 10*

actions or the MATLAB >> prompts you see on your screen.

Bold type is used for the letters and numbers to be input by you.

Bold italic type indicates actions you are to take.

Sans Serif Font shows text you see on your computer screen, for example, messages and command prompts.

Here are some examples:

>> ***format short [Enter]***

instructs you to type "format short" then press [Enter] in the Command window where you see the >> prompt. This will execute the format command with the short option. Once the command is executed, a new prompt appears.

The symbol >> represents the MATLAB command prompt that you see in the Command Window on your MATLAB application screen.

Instruction words, such as "*Click*," "*Type*," "*Select*," and "*Press*," also appear in sans serif font and are italicized. For example, "*Type*" instructs you to press several keys in sequence. For the following instruction,

Type: ***4.2315 - 3.2145***

you would type 4.2315 - 3.2145 in that order on your keyboard.

"*Click*" tells you to click an icon or to choose commands from the menu. For example,

Click: ***PLOTS tab***

instructs you to select the PLOTS tab from the ribbon menu near the top of the application by clicking with the mouse. To select a tab or icon, move the pointer until the cursor is over that tab or icon and click the mouse button.

An Instruction shown like this:

Press: ***[Enter]***

means you should press the Enter key on your keyboard once.

Sometimes "press" is followed by two keys, such as

Press: ***[Ctrl]+[F1]***

In this case, press and hold down the first key, press the second key once and release it, and then release the first key.

You will sometimes be instructed to perform some steps on your own. These instructions will be listed in a different font like this:

On your own, clear the command window using the clc command.

End-of-Tutorial Exercises

Exercises at the end of the tutorials provide practice for you to apply what you have learned during the tutorial. To become proficient at using MATLAB, it is important to gain independent practice by trying some of the exercises..

Creating a Working Directory

To make saving and opening files easier as you complete the tutorials, you will create and specify a new working directory to use with MATLAB. A directory is another name for a folder. For example, you may have a document file, *example.doc,* and store it in a folder (directory) called *MyDocuments*.

The *working directory* is the default directory used in the application to open and save files. By setting a working directory, you will not have to search through the computer's directories every time you are asked to save a file.

This is a good practice in general for organizing your files. You should not save your MATLAB files and other work files in the same directory as the MATLAB software because doing so makes it likely that they will be overwritten or lost when you install software upgrades. It is also easier to organize and back up your project files and drawings if they are in a separate directory. Making subdirectories within this directory for different projects will organize your drawings even further.

You will create a new shortcut for launching the MATLAB software that will use *c:\Work* as the starting directory. If you are used to keeping your files in the My Documents folder on your computer, then make a new directory named *Work* inside your My Documents folder.

> **On your own, use your Windows Explorer to create a new directory,** *c:\Work* **at this time.**

When you are finished, return to the Windows desktop.

Installing Data Files for the Tutorials

The next step is to create a directory for the data files used in the tutorials and install them. Throughout the tutorials, you will be instructed to work with files already prepared for you. You will not be able to complete the tutorials without these files. You can download these files with a Web browser such as Chrome, Mozilla Firefox or Internet Explorer.

> **Navigate to** *www.sdcpublications.com* **using your web browser.**

Once you are at the site,

> Click: **Authors tab,** *then choose letter* **L**
>
> *From the author list* **choose Lockhart, Shawna** *to show this and other books*
>
> **Choose your text by clicking on the picture of its cover**
>
> *From the book's page, choose the* **Download tab**
>
> Click: **Download File button**
>
> **Follow the instructions to download the data files**

After you have retrieved the files, follow the next steps to extract the data files to a new directory.

Warning: *Changes to the MATLAB software's configuration can affect its performance greatly. If you are working on a networked system, you must check with the network administrator before making configuration changes.*

Use your Windows File Explorer (or other operating system tools if not Windows) to create a new folder named dataf iles-matlab.

Figure GS1.1 shows an example of right-clicking in the Windows File Explorer to select New > Folder.

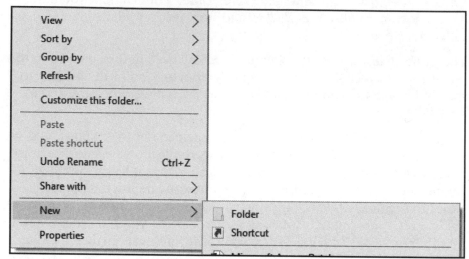

View	>
Sort by	>
Group by	>
Refresh	
Customize this folder...	
Paste	
Paste shortcut	
Undo Rename	Ctrl+Z
Share with	>
New	>
Properties	

Folder
Shortcut

Figure GS1.1   Right-Click Menu from Windows File Explorer

Use your operating system tools to copy dataf iles-matlab.zip into the newly created directory.

Right-click: dataf iles-matlab.zip file and select Extract All... or use a utility such as WinZip to extract the zip file.

☑ dataf iles-matlab.zip

Open
Open in new window

Extract All...

Figure GS1.2   Right-click Menu for Extract All

When the files are extracted, you will see an assortment of files in the directory, some of which have the .m extension.

Configuring MATLAB for the Tutorials

You will now set up your system so that you can work through the tutorials as instructed.

To start the MATLAB software,

Click: MATLAB shortcut from the desktop

The MATLAB graphics window appears as shown in Figure GS1.3.

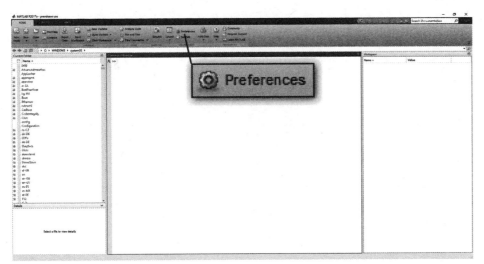

Figure GS1.3   MATLAB Software Window

*Click: **Preferences** from the center of the ribbon.*

The Preferences dialog box appears on your screen.

*Click: **Current Folder** from the topics at the left of the dialog box*

The options for the folder preferences show in the dialog box similar to Figure GS1.5.

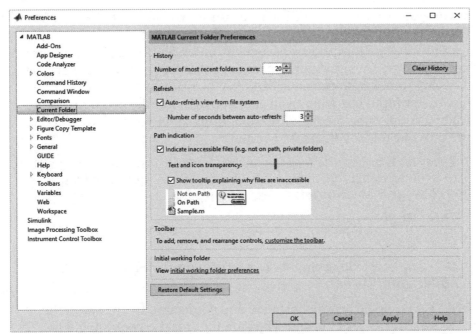

Figure GS1.4   Current Folder Preferences Dialog Box

*Click: **initial working folder preferences** from near the bottom of the dialog box*

The Options dialog box appears on your screen, as shown in Figure GS1.5.

The General Preferences page of the dialog box lets you set up the default working directory.

Note: *If you are not going to use c:\Work for saving your files, you can set it to the folder name you will be using. If not, leave the setting to the Last working folder from the previous session.*

Tip: *Notice that many of the pages in the Preferences dialog have a Restore Default Settings button. If you make unwanted changes, try restoring the defaults.*

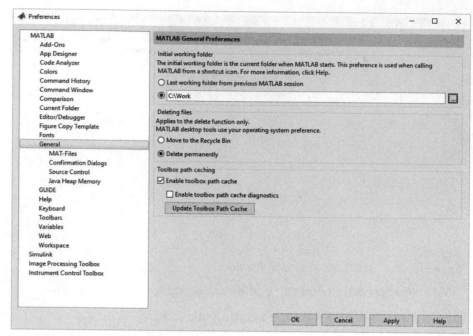

Figure GS1.5   General Preferences

Initial Working Folder

When you launch the MATLAB software the Current Folder window within the interface will display either the last folder in use, or you can set the software to start in a particular folder. Notice the buttons near the top of the General Preferences page that let you select either Last working folder from previous MATLAB session, or the other option next to an input box where you can type or select the default working folder. You use the second of these options to set your software to use the *c:\Work* folder you created earlier.

Type: **c:\Work** *as the initial working folder or use the ... (ellipsis) to browse to the folder location where you will store the files you create for these tutorials*

Click: **Apply**

Click: **OK**

On your own, close the MATLAB software using the X in its upper right corner

MATLAB should now be installed and configured to open with the *c:\Work* directory set current. The data files for the book should be unzipped and available for use in a folder named *c\datafiles*-matlab. You have completed this section.

INTRODUCTION TO MATLAB®

Introduction

You have MATLAB installed, licensed, and ready to use, so let's use it! This tutorial introduces the fundamentals of MATLAB 2017. It briefly explains how to enter mathematical expressions and work with variables. You will learn how to use the help system and how to store variables in a file for later loading. You will also learn some handy editing features. This is a quick overview and we will come back to these topics in later tutorials in more detail.

Starting

Launch the MATLAB 2017 software.

If you need assistance, refer to the Getting Started chapter.

The MATLAB Screen

Figure 1.1 shows the MATLAB application. The default environment should be open on your screen. The MATLAB desktop shows the title bar across the very top with the Windows controls for closing, minimizing, and maximizing the application at the upper right of the application window. A quick-access toolbar just below the title bar has icons for help, and a search box. The ribbon contains tabs for HOME, PLOTS, and APPS. The HOME tab displays by default.

The central panel is the Command Window, where you enter MATLAB commands by typing them at the >> prompt. The Workspace area is at the right by default. This area lets you explore variables and functions you have in use. The left panel shows your current folder and allows you to access files.

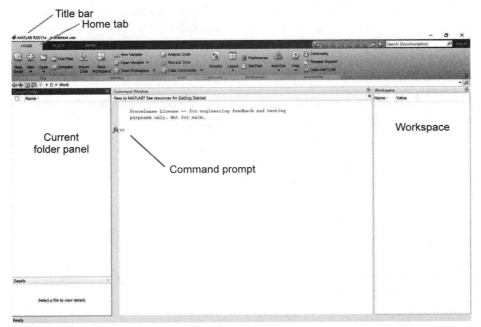

Figure 1.1   **The MATLAB Application Desktop**

Objectives

When you have completed this tutorial, you will be able to

1. Identify parts of the MATLAB screen.

2. Enter a math expression.

3. Clear the Workspace and Command Window.

4. Work with variables.

5. Use the Help command.

6. Save and load Workspace variables.

7. Show the command history.

8. Use a built in function.

Tip: *If your screen does not look generally like Figure 1.1 you may have settings that are not typical defaults. Refer to the Getting Started section of this manual for more information. This book assumes you are using the Windows desktop version of the software. Other versions are similar, but will not match this text exactly.*

Tip: *You can resize and rearrange the panels by clicking and dragging on their dividing borders.*

Entering Commands

Commands are entered at the >> *prompt* in the Command Window. Pressing [Enter] causes the result to be displayed. Try it now.

>> *126 + 328*

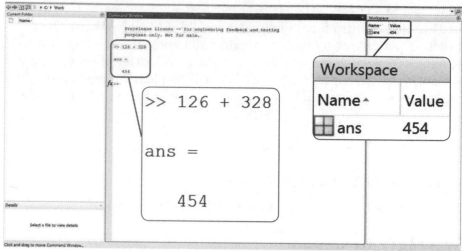

Figure 1.2   **The Command Window**

Notice the result. MATLAB displays **ans = 454**. In the Workspace window a *variable* named *ans* was created which now has the value 454. When you enter an arithmetic expression but you have not assigned a variable, one named *ans* is automatically created for you. Try another.

>> *532 - 33.9123456*

MATLAB displays **ans = 498.0877**. A new value has been assigned to the variable, *ans*, in the Workspace.

Notice that the decimal places are rounded from 498.0876544 to 498.0877 in the display of the result. Values in the computer's memory are stored with 16 digits of precision even when the display format is set to short format which shows only four decimal places. You can change the format for the display from the command prompt or by using the Preferences dialog box. When necessary, you can increase the internal accuracy with which values are stored, but this may slow processing for complex calculations.

You will learn more about operators in the next chapter, but for now, keep these in mind. They probably seem familiar.

+   addition (plus)
-      subtraction (minus)
*   multiplication (times)
/   division (divide)
^   power

Preferences

The Preferences dialog box lets you quickly set the default behaviors for the MATLAB software.

Click: **Preferences** *from the ribbon HOME tab*

The Preferences dialog box appears as shown in Figure 1.3.

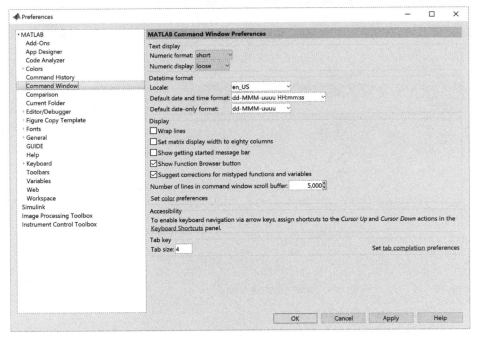

Figure 1.3   **The Preferences Dialog Box**

The left side of the dialog box shows a list of topics. Preferences for the selected topic, in this case Command Window, appear in the dialog box.

> *Select: **long** from the drop-down list for Numeric format.*
>
> *Click: **Apply***
>
> *Click: **OK***

Try it out by dividing one by three.

> \>> *1/3*

MATLAB now displays 16 digits. You should see
ans = 0.333333333333333 in the Command Window.

> *On your own, use the Preferences dialog box to **set Numeric format to short** before continuing.*

Using Help

MATLAB has an excellent help system. MATLAB has many commands, libraries, options, and toolboxes. When using MATLAB, commands must be entered exactly. The help system is invaluable in learning commands, looking up command formats, and exploring the available tools.

> *Click: **Help** to expand the menu options shown in Figure 1.4.*

In addition to the documentation, there are also excellent examples, learning tools, support, and a community of users.

 *Click: **Help icon***

Figure 1.5 shows the Help window. The search box is located at the upper right of the window.

Tip: *Use the* format *command followed by one of the following options:* short, long, shortE, longE, shortG, longG, shortEng, longEng, +, bank, hex, *or* rat *to set the digit display using the command prompt.*
Eg: >> format long

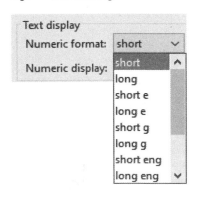

Tip: *You can set the numeric display to compact or loose. This controls the line spacing in the Command Window. Compact display is often useful.*

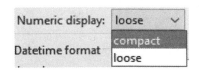

Figure 1.4
Help Selections

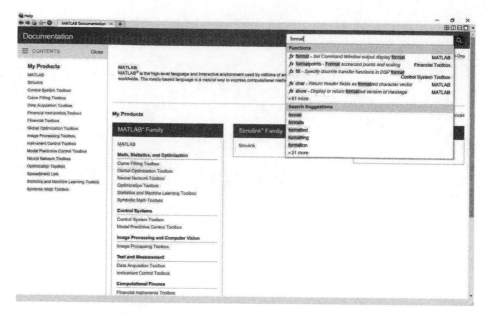

Figure 1.5   **Help Browser Search Box**

Type: **Format** *into the search box*

Matched items appear as you type into the search box.

Click: **fx format - Set Command Window output display format**

Scroll down to show the additional content

Figure 1.6   **Results for Format Command**

Examine the different options available for the output display. Get used to using the help system. You can learn a lot from browsing the help.

Click: **X** *to close the help window*

You can also enter the help command at the >> prompt. Give it a try.

>> *help*

Use the scroll bars at the right of the Command Window to access the top of the list as shown in Figure 1.7.

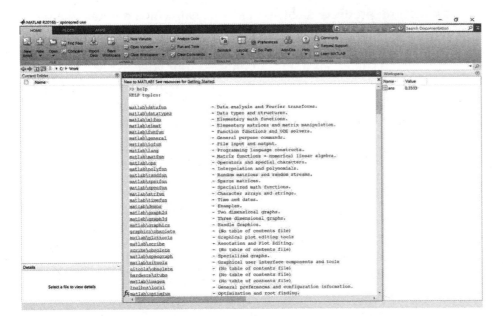

Figure 1.7 **Help Results in the Command Window**

Click: matlab\general - General purpose commands

On your own, look over the list of general commands

Clearing the Command Window

The **clc** command clears the Command Window. Try it now.

>> *clc*

The Command Window should now be blank.

Clearing the Workspace

The clear command clears the Workspace variables. Go ahead, try it.

>> *clear*

The Workspace should now be blank. Next, you will use MATLAB to convert some data from a file. Make sure you have the data files for this book loaded. See the Getting Started section for information on how to download the files.

Using MATLAB: A Unit Conversion Example

Units are an important part of calculations. From Christopher Columbus mistaking the size of the earth due to the difference in European and Arabic miles of that era to the Mars Climate Orbiter disintegrating when approaching too close to the planet, resulting from data provided in US customary units when the NASA scientists specified SI (metric) units, errors resulting from incorrect units cause serious mishaps.

The next project will be to use MATLAB to convert values stored in a data file for use with different units. During this process you will learn how to open and save to a .mat file. You will also begin working with variables.

Tip: *A little overwhelmed at all of these commands? Some of these like* clear *and* clc *you will use frequently. Typically you will use help to look up formatting and command details. You are not going to memorize all or even most of the commands.*

Tip: *You can also select the* clc *command from the ribbon menu. Notice the downward triangle which you can click to expand additional options.*

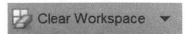

Tip: *The* clear *command is also available from the ribbon.*

A manufacturer produces a chromium steel ball with a nominal diameter of 25 mm. They automatically measure the ball weights to inspect for manufacturing defects. A sample of their inspection data is stored in ballbearing_data.mat. This data is in ounces. Your goal: use MATLAB to convert the data to grams and inspect the data to find the average mass.

Information
Ball Bearing Diameter: 25 mm
Material: ASI 52100 Chromium Steel
Desired Ball Weight (each): 66.9500 grams ± 0.12

Loading a File

A strength of the MATLAB software is the ability to import and work with large data files from many formats, including spreadsheets, text, and image files. For this project, you will open a small MATLAB data file. You will work with importing spreadsheet data in a later tutorial.

*Click: **small sideways arrow** (it turns downward) to expand the folders list and select **datafiles-MATLAB** to make it the current folder or use the Browse for folders button to show the file browser.*

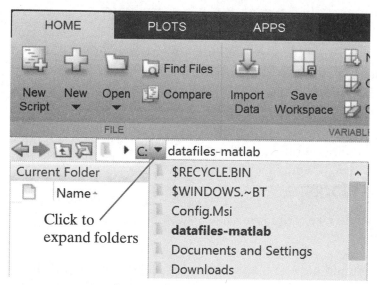

Figure 1.8   **Browsing to a Folder**

*Double-click: **bearingdata.mat** from the list of file names.*

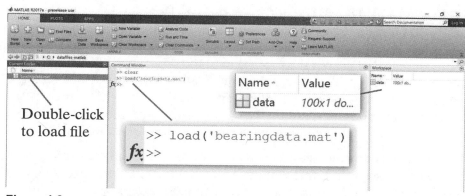

Figure 1.9   **Loading a File**

Tip: *Make sure you have downloaded and unzipped the data files for this text. See the Getting Started section at the beginning of the book for details..*

Tip: *You can select the Browse for folders icon to open a file browser to locate the data file.*

Double-clicking on the file name *bearingdata.mat* automatically loaded it. Notice the `load` command was executed in the Command Window (Figure 1.9). You can also enter `load ('filename.ext')` at the >> prompt, of course replacing *filename.ext* with your actual filename and file extension (*.mat* for MATLAB data files, *.m* for scripts).

In the Workspace, the variable named *Data* is now loaded.

A First Look at Variables

MATLAB variables are named storage locations to which values may be assigned. Variables must start with a letter and may include numbers and the underscore character (_). It is a good practice to use variable names that provide enough description that you can identify them easily. Variable names are *case-sensitive!* For example, kwh2 is an allowed variable name, but Kwh2 is not the same variable. 2kwh is not an allowable variable name. You cannot use variable names that are the same as MATLAB keywords (such as " if" or "end"). You should also take care not to name your variables the same as a built-in MATLAB function. This is allowed, but it may produce unwanted results.

The variable Data, which is now loaded, is an *array* of 100 elements in one column. All MATLAB variables are forms of an array or matrix. A *matrix* (the plural is matrices) is essentially a rectangular grid of elements. Even variables which contain a single value are a special case of a matrix, where there is only one column and one element in it. You'll learn about arrays and matrices in-depth in a later tutorial.

MATLAB is a shortened version of Matrix Laboratory. Ease in working with matrices is what sets MATLAB apart from spreadsheets and equation-solving software you may be familiar with. While the introductory examples in this book may be easy to do using a spreadsheet, you will be learning to use a powerful tool with features for manipulating matrices and working with large data sets that are not easily accomplished with other tools.

Viewing the Data Variable Contents

To show the elements of the variable named *data*, simply type its name at the >> prompt.

> **>> data**

You should see a list of 100 numbers, such as the final ones 2.3633, 2.3637, and 2.3629. These values are in ounces. Your goal is to convert the ounce weight values to grams of mass. One ounce of weight measured at standard gravity (9.81 m/s^2) has a mass of 28.35 grams.

> **>> data_grams = data * 28.35**

In the Command Window, the 100 values have each been multiplied by 28.35 to result in 100 numbers ending in 66.9987, 67.0122, and 66.9887.

Next, let's save the new values as a *.mat* file so it is available to load in the future if needed. Anything you have not saved to a file of some sort is gone forever the next time you clear the Workspace. You can clear the Command Window without clearing the Workspace.

Tip: *Notice that the load command expects the filename given in the parentheses to be enclosed in single quotes. This is not typical of every programming language.*

Note: *The following MATLAB keywords cannot be used as variable names:*

> break
> case
> catch
> classdef
> continue
> else
> elseif
> end
> for
> function
> global
> if
> otherwise
> parfor
> persistent
> return
> spmd
> switch
> try
> while

Note: *1 ounce equals approximately 0.278 Newtons. A gram is a unit of mass, while the Newton and ounce are units used in measuring weight. A gram of material has that same mass on Earth as on the Moon, but its weight in ounces (or Newtons) will be different on the Earth and the Moon. Weight here is the measure of gravitational force exerted on the object by the Earth.*

Tip: save ('filename.ext','variable') *is the function form of the command. You can use the simpler command form which does not require the parentheses or single quotes, but this works only when entered directly at the >>prompt and not inside scripts or functions, which you will be learning in later chapters.*

Tip: *If you had trouble saving your file, it may be that you are not able to write to the datafiles-matlab folder. When you are using files in a networked environment, the folder may be read-only to prevent the data files from being accidentally changed. If this is the case, use* *the Browse for folder icon and select your c:\Work folder or another folder where you have permissions to create files. You can then use the up arrow to show the command history and repeat the save command with that current folder in use.*

Tip: *If you are using the MATLAB Online version, you can use the up arrow to scroll through previous commands, but you won't see the history window.*

Saving a Variable to a .mat File

Use the save command to save a variable to a file. The general form of the command is save('filename.ext','variable'). You will name the new file bearingdata_grams and use your data_grams variable as the one to save. The .mat extension is provided automatically, or you can enter it with the filename.

> **>> *save ('bearingdata_grams', 'data_grams')***

You will see the new file listed in your current directory, shown in the Current Folder window, similar to Figure 1.10.

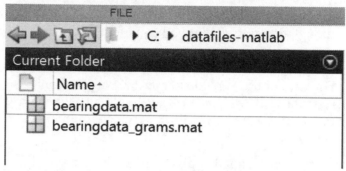

Figure 1.10　　　**Current Folder with bearingdata_grams.mat**

Showing the Command History

You can press the up arrow (↑) on your keyboard to show the command history. This is an easy way to review what you have entered, select an item to quickly edit typing errors, or to repeat a similar command.

> *Press:* ↑

The command history is listed above the >> prompt in the Command Window, similar to Figure 1.11. You can resize the window and scroll up or down to select entries. When an item is selected, it appears at the >> prompt, where you can make changes. This is great is you have forgotten a parenthesis or made a typing error.

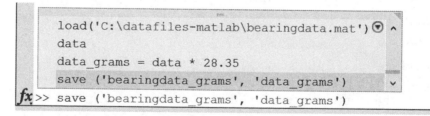

Figure 1.11　　　**Command History**

Working with Built-in Functions

Built-in functions allow you to do many useful operations quickly. Next, you will use some of the statistical functions to find information about the ball bearing data you have saved.

Basic Statistical Functions in MATLAB

min	Smallest elements in array
max	Largest elements in array
bounds	Smallest and largest elements
mean	Average or mean value of array
median	Median value of array
mode	Most frequent values in array
std	Standard deviation
var	Variance
corrcoef	Correlation coefficients
cov	Covariance

Let's find the lowest ball bearing weight in the sample data.

> **>> min (data_grams)**

You should see: ans = 66.9075.

> *Press:* ↑ *and edit the previous line from min to max and then press [Enter]. You can highlight the word "min" with your mouse cursor and overtype it or use most typical editing methods*

> **>> max (data_grams)**

When you press [Enter] you should see: ans = 67.0477.

> **>> mean (data_grams)**

ans = 66.9709

> **>> median (data_grams)**

ans = 66.9721

> **>> [smallest,largest] = bounds (data_grams)**

smallest = 66.9075, largest = 67.0477

The bounds command created two new variables (smallest and largest). The function assigned the first result (the smallest element) to the first variable in the list to the left of the equal sign, and the second result (the largest element) to the next variable in the list. Notice that these variables are shown in the Workspace now.

> **On your own, clear the Workspace and the Command Window.**

In the next section, you will create a script you can reuse for solving some basic physics of falling objects.

Terminal Velocity Example

Think about what would happen if you were in an airplane and dropped a steel ball out of the window. Okay, we get it that other bad things would be happening and you wouldn't be here reading this, but just imagine for the sake of example. An object falling through the Earth's atmosphere encounters air-resistance effects. At some point the object may reach terminal velocity. Terminal velocity is the highest velocity an

? What is the difference between a mean and a median?

The mean is the average of the values (total/count). The median is the dividing value between the upper and lower elements of the sample, if the elements were sorted into ascending or descending order and you found the middle item. When there are an even number of elements in the array, the median is the two middle values divided by two.

object can reach as it falls through a medium (in our example, the air). This results from the drag force (the friction effect of the air) on the object, which increases as the object's velocity increases. To calculate the terminal velocity we will use equation (1), where:

Vt = terminal velocity

m = mass of the falling object

g = acceleration due to gravity (approximately 9.81 m/s^2)

ρ = density of the medium the object is falling through (which we will take as air with a density of 1.225 g/m^3).

A = cross-sectional area of the object (as if you projected it onto a plane perpendicular to the direction of motion, sort of like the silhouette of the object when looking from below, a circle in the case of a sphere).

C = drag coefficient. The drag coefficient depends on the shape of the object. More streamlined shapes have lower drag coefficients. Factors such as surface roughness influence the drag coefficient. These are often determined experimentally. The drag coefficient for a sphere is approximately 0.47.

Terminal velocity equation:

$$Vt = \sqrt{(2mg)/(\rho CA)} \tag{1}$$

Creating a Script

Scripts are a quick way of storing commands so that you can run them again later. There are two types of scripts in MATLAB, scripts and Live Scripts. In this section you will be using a script, not a Live Script. When you use the edit command with a file name, the file is automatically created. Watch out, this may make it somewhat easy to accidentally write over an existing file.

> *On your own, set C:\Work as your current MATLAB folder.*

> *>> edit terminal_velocity*

You will see a message asking whether you want to create the file (Figure 1.12). Notice that the file is being created in the current working directory. Scripts have the .m file extension.

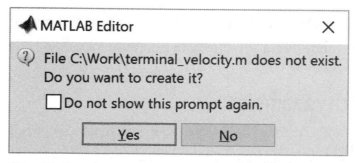

Figure 1.12

*Click: **Yes** to create the terminal_velocity.m file*

Note: *As an object falls through the atmosphere, the air density changes with the altitude and the air temperature. At 15 °C and sea level, air has a density of approximately 0.001225 g/cm3 according to ISA (International Standard Atmosphere).*

The link http://www. engineeringtoolbox.com/drag-coefficient-d_627.html has a chart of drag coefficients for some fun shapes, if you want to see the difference between dropping a dolphin instead of a spherical steel ball.

Tip: *You can use the New Script selection from the ribbon Home tab to create a new script. If you do this remember to save the script with a name later.*

The Editor window opens on your screen as shown in Figure 1.13.

Figure 1.13   The Editor Window

On your own, use the Editor window to enter the following lines:

% program to calculate terminal velocity

% Vt = terminal velocity

% dia = diameter of sphere (in cm)

% mass = mass of the falling object (65.4710 grams)

% gravity = acceleration due to gravity (about 981 cm/s^2)

% airDensity = density of the medium (0.001225 g/cm^3.)

% area = cross-sectional area of the object (in cm^2),

%   which for a sphere is pi∗radius^2

% C = drag coefficient (about 0.47 for a sphere)

dia = 2.5;

area =. pi∗((dia/2)^2);

mass = 65.4710;

gravity = 981;

airDensity = 0.001225;

C = 0.47;

Vt = sqrt((2∗mass∗gravity)/(airDensity∗C∗area))

disp 'cm/s units'

Notice the various colors used to help you keep track of things in the Editor window. For example, the comment statements are green on the screen. You can use the Preferences dialog box available from the ribbon Home tab to select different colors or to check what is being indicated by this color coding. See Figure 1.14.

Tip: *% indicates a comment line, so the item appearing afterward is not executed when the script is run in the Command Window.*

Tip: *A semi-colon (;) at the end of a line suppresses the output so the result of that line is not displayed in the Command Window.*

Tip: pi *and* sqrt *are built-in functions. You will learn more about functions in a later tutorial.*

Note: *If you are color blind, you may want to choose colors you can easily distinguish, instead of the default green for comments and red for warnings.*

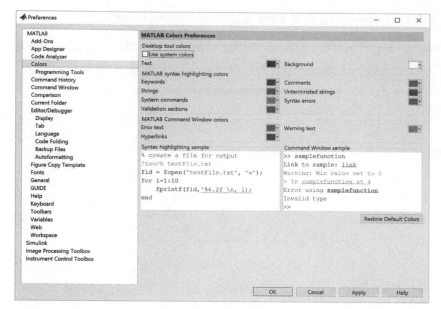

Figure 1.14   The Preferences Dialog Box

The *Code Analyzer* is a handy tool when creating scripts, functions and other code. You can quickly select it by expanding the context menu at the upper right of the Editor window. See Figure 1.15.

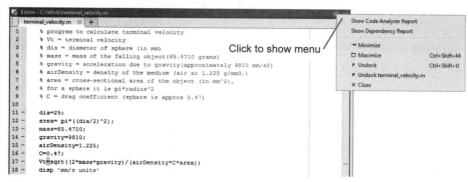

Figure 1.15   Editor Window with Context Menu

Click to expand the menu and select ***Show Code Analyzer Report***

The Code Analyzer Report has useful debugging information and tips on improving your MATLAB code. Use it thoughtfully. Close the Code Analyzer Report when you have finished reviewing the information.

Figure 1.16   Code Analyzer Report

Now you are ready to have a try at running your script.

Running a Script

To run a script, you can select the Run button from the ribbon's Editor tab or you can type the script name into the Command Window.

On your own, run the script terminal_velocity.

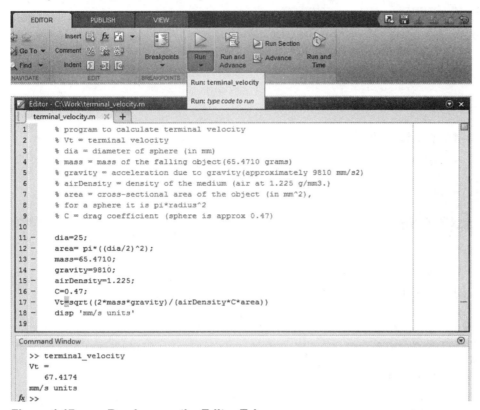

Figure 1.17   Run Icon on the Editor Tab

Your result in the Command Window appears similar to:

Vt =

67.4174

mm/s units

You will learn more about writing code and formatting your output in upcoming tutorials.

Saving Your Files

Use the operating system to save your files. If you are using a shared computer network, make sure that you have copied your files into a folder that you can access later. Copy the files you created in this tutorial onto a removable storage device, such as a USB drive.

On your own, copy the file to the drive where you want to save the file.

Closing MATLAB

Use the Windows close box to exit the MATLAB software. You have completed the first tutorial!

Tip: *It is extremely important to keep track of units in calculations. Adding labels to the input and output formats, automatically checking that input values are within the allowed range, and using variable names and prompts that request the value in particular units are all helpful ways to prevent errors.*

Key Terms

>>	*array*	*comment*	*variable*
%	*arithmetic operators*	*matrix (matrices)*	*Workspace*
;	*command prompt*	*script*	
ans	*Command Window*	*suppress display*	

Key Commands

clear   format style

clc   help

Exercises

Exercise 1.1

A climbing carabiner is rated to have a closed gate strength of 20 kN (kiloNewton) and open gate strength of 6 kN. Use MATLAB to calculate the rating for the carabiner in lbf (pounds force). 1 kN = 224.809 lbf.

Exercise 1.2

Use the data file *bearingdata.mat* to convert the values stored in the file from ounces to kilograms. One ounce equals 0.0283495 kg. Show the values to 16 significant digits.

Exercise 1.3

Use MATLAB to calculate the following:
  a. 654.23 + 345.73
  b. 597.64533 - 0.83455
  c. 3984566 / 5.5
  d. 343490999 / 2 * 34.6540

Exercise 1.4

Modify the terminal_velocity script you created previously to calculate the terminal velocity if the 25 mm spherical object of mass 65.4710 was falling towards Mars and not Earth. The gravity of Mars is 3.711 m/s^2 . The atmospheric density at the surface of Mars is estimated to be .020 kg/m^3. Notice that these units for the density and gravity of Mars are in m (meters) units not mm (millimeters).

PROGRAMMING BASICS: OPERATORS & VARIABLES

Introduction

This tutorial gets you started working with MATLAB as a programming language. There are a whopping number of programming languages out there, but a lot of them use the same handful of basic elements. This means that after you learn your first programming language, you've got a conceptual leg up on many of the rest and only need to learn new syntax (the "grammar" rules) to hit the ground running. The basic concepts will be familiar if you have done a bit of programming, but if so, look for MATLAB-specific syntax and usage information.

Operators

Operator is a fancy name for symbols like +, -, and <. Operators are normally broken into usage-based categories: arithmetic, relational, and logical. More operators will be discussed when we get to matrices.

Arithmetic Operators

Arithmetic operators are used on numbers and return a number. You used them in our earlier tutorial and they are similar to functions on a calculator, so these should seem familiar.

Operator	Action
+	Addition (3 + 2)
-	Subtraction (3 – 2)
*	Multiplication (3 * 2)
/	Division (3 / 2)
^	Raise to the power (3 ^ 2)

Relational Operators

Relational operators make a comparison. The comparison results in a *logical value*. A logical value is one that has just two states: true or false, which are represented by 1 for true and 0 for false.

Operator	Meaning
<	Less than
<=	Less than or equal to
>	Greater than
>=	Greater than or equal to
==	Equals
~=	Does not equal

Objectives

When you have completed this tutorial, you will be able to

1. Understand and apply operator precedence.

2. Assign a value to a variable.

3. Understand the minimum and maximum value of various numeric variable types.

4. Understand the difference between string and character arrays.

5. Work with data from an imported file.

Note: *MATLAB is the name of both the programming environment (the MATLAB program itself) and the programming language used in that environment. Both uses are fairly interchangeable in conversation, if you ever have those sorts of conversations.*

Tip: *Operators also have function forms; for instance* plus(3,4), *which does the same thing as 3 + 4. Their reason for being is beyond our scope and we won't be using them. See the MATLAB help section on operators if you're curious to learn more.*

Not All Equals are Equal

A quick word about the *equals sign operators*. MATLAB has more than one operator that uses equals signs (=) and they act differently. If you have done some programming, you are probably used to the idea that "=" is not used for comparison, but rather to *assign* the value from the right side of the equals sign to the variable on the left side. For example:

index = index + 1

is a perfectly legitimate *assignment statement* (for the new value, add 1 to the old value), but the two sides are certainly not mathematically equal.

Type the following lines in the Command Window and observe the results:

>> x = 3

>> x == 3

The results display in the Command Window similar to Figure 2.1.

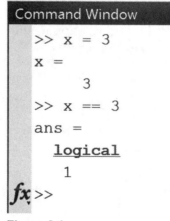

Figure 2.1

The command x = 3 assigns the value 3 to the variable x. The command x == 3 asks "Is x equal to 3?" and returns a value of 1, which means "true" since we just set it equal to 3. We'll talk about this again, but the accidental use of = when you meant == is a common coding error.

Logical Operators

Here are the *logical operators*. These operators are used on logical values and result in a logical value.

Operator	Meaning
&&	and
||	or
~	not

Logical operators might be new to you. Logical operators test for true or false and report either 1 (true) or 0 (false). In your code, you can either use 1 and 0 or the keywords **true** and **false**.

The not (~) Operator

The (**~**) operator (not) is straightforward: it reverses the logical value of the expression that it's applied to, so that true (1) becomes false (0) and false (0) becomes true (1). You can use either the symbol **~** or the command word, **not**.

Type this line into the Command Window for a demonstration:

>> ~ (x == 3)

Since x was previously set to 3, the value within the parentheses is evaluated is true (x is exactly equal to 3), then the ~ operator is applied to reverse it to false (0). The Command Window displays: **ans = logical 0**.

The and (&&) Operator

The **&& operator** (**and**) returns true only when *both* of the logical values it compares are true. This grid shows the results of **X && Y** for logical values X and Y:

X	Y	X and Y	Explanation
true (1)	true (1)	true (1)	X && Y is true if X is true and Y is true
true (1)	false (0)	false (0)	X && Y is false if X is true and Y is false
false (0)	true (1)	false (0)	X && Y is false if X is false and Y is true
false (0)	false (0)	false (0)	X && Y is false if X is false and Y is false

The or (|) Operator

The **|** **operator** (**or**) returns true when *either or both* of the logical values it compares are true. This grid shows the results of **X | Y** for logical values X and Y:

X	Y	X or Y	Explanation
true (1)	true (1)	true (1)	X | Y is true if X is true and Y is true
true (1)	false (0)	true (1)	X | Y is true if X is true and Y is false
false (0)	true (1)	true (1)	X | Y is true if X is false and Y is true
false (0)	false (0)	false (0)	X | Y is true if X is false and Y is false

The && and | operators compare expressions that return a logical value rather than comparing simple logical values. We'll scc a lot of this when we talk about *if-then* and *while* statements.

For now, experiment a little by entering the next lines in the Command Window and noting the results.

Tip: *The tilde (~) is usually at the very upper left of the numbers row on your keyboard.*

Tip: *Want to get rid of a single variable from your Workspace without clearing them all? Use the clear command as usual, but specify the variable to delete. For example,* clear x.

Tip: *You can type && or you can use the command word* and. *You can type | or use the command word* or.

Tip: *Exclusive* **or** *means A* **or** *B, but not A* **and** *B. This operator is available as* xor.

Tip: *The* **or** *operator has two versions that are not the same: | and ||. The element-wise* **or** *operator is |. The* **short-circuit or** *operator is ||.*

Tip: *Check the Workspace to see the value your variable has. If the Value column doesn't show, right-click to the left of the Name area and use the menu to select Value.*

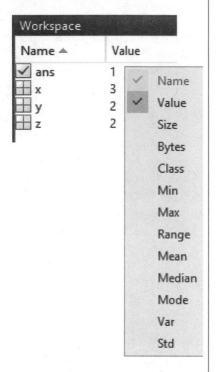

Tip: *Remember the semicolon (;) suppresses the output from the entered line.*

Tip: *Functions generally have some input or "argument" that they act on. For example:*
        islogical (x)
asks is the "argument" x of the type logical. If it is the answer is 1 (true), if not the answer is 0 (false). We write islogical() to show you that the function expects an argument to be entered in the parentheses when you use this function. You will be learning more about functions in the next tutorial.

Work through each side of the logical equations to understand why each line returns the logical 1 (true) or 0 (false) that it does.

Command Window entry	Result
x = 3;	assigns value of 3 to x
y = 2;	assigns value of 2 to y
z = 2;	assigns value of 2 to z
x == y	ans = logical 0 (false)
y == z	ans = logical 1 (true)
x == y && y == z	ans = logical 0 (false)
x == y | y == z	ans = logical 1 (true)
~(x == y) && y == z	ans = logical 1 (true)

The Logical Data Type

Relational and logical operators return a variable that is of type *logical*. This type has two values, 0 for false and 1 for true. Logical 1s and 0s are not the same as normal 1s and 0s, but the differences can be subtle. Watch out for situations where a statement expects a logical value. Though MATLAB will often automatically do any necessary conversion for you, a good programmer doesn't count on that. When a logical value is needed in your code, use the built-in MATLAB function logical() to create it. Type these lines into the Command Window and note the type that x and y are assigned.

>> *x = 1;*

>> *y = logical(1);*

>> *class(x)*

You see the result, ans = 'double'. Double is the default numerical data type in MATLAB. It stores values between ± 3.4 x 10^{38}.

>> *class(y)*

You see the result, ans = 'logical'.

To check whether a value is of type logical, use the MATLAB function islogical(). Type these lines into the Command Window:

>> *islogical(x)*

>> *islogical(y)*

While MATLAB returns the logical 1 and 0 in the Command Window, in your code you can also use the MATLAB constants true and false, which are the same as logical(1) and logical(0).

Operator Precedence

Operator precedence is the order in which operations are executed in a statement that contains more than one operator. We just saw some examples above. The statements x == y and y == z were executed before the && or || operator was used. This shows that the == relational operator has a higher precedence than the logical && and || operators.

Operator precedence is vital to know, because the result of an operation can be wildly different than intended if precedence isn't understood and addressed. For instance, enter the following two lines into the Command Window and note the different results:

```
>> 3 + 2 * 6
>> (3 + 2) * 6
```

Multiplication has a higher precedence than addition, so the first line is equivalent to 3 + (2 * 6). Parentheses override precedence, so the second line first adds 3 + 2 and then multiplies the result by 6. Someone who understands the precedence rules might use line one, but someone who assumes that everything is done left-to-right might use the first when they meant the second. Use parentheses to ensure the correct precedence!

Order of Operations

1. Parentheses
2. Logical negation (~), unary minus (-)
3. Multiplication, division
4. Addition, subtraction
5. Relational operators (<, <=, >, >=, ==, ~=)
6. Logical AND (&&)
7. Logical OR (||)

In the case of a tie, operations are usually executed from left to right. There are many more operators and you can see the full list by searching for "operator precedence" in the MATLAB help. Scoff now, but if you do enough programming, you will know the full list by heart one day.

Variables

The concept of a *variable* is straightforward: a variable holds a value. It has a name, such as *x* or *city*, and that name is assigned a value, such as 3 or 'Paris'.

Variable Naming Rules

As discussed previously, variable names are case-sensitive (abc and Abc are different variables), they can use letters, numbers, and the underscore character ('_') and must begin with a letter. They can be no longer than 63 characters!

Enter the following commands into the MATLAB Command Window and observe, in particular, the Workspace window. The Workspace window provides an at-a-glance list of all active variables you've defined, with their values. Some of these commands will cause an error. Don't be alarmed.

```
>> a = 5
>> A = 15
>> a + 3
>> A + 3
>> A + a
```

Tip: *You may remember the mnemonic My Dear Aunt Sally (multiply divide add subtract) from elementary school for the order of arithmetic operations. MATLAB evaluates arithmetic operations in that same order.*

Tip: *A unary minus is what you probably know as a negative sign; for example, -3. "Unary" means that it only has one operand.*

Tip: *Technically, a variable's name is an alias for a location in memory where that value is stored. Sometimes that is helpful to remember.*

Tip: *When you provide an invalid variable name, MATLAB's error message doesn't specifically say that's the problem. Look for errors that talk about the value to the left of the equals sign, though there are others as well.*

Tip: *MATLAB ignores most white space, so a=3 and a = 3 are both valid, as is a+3 or a + 3.*

Note: *Did you notice that when you entered an invalid statement, MATLAB tried to help you out by displaying "Did you mean:" with a suggested alternative? If that alternative is correct, just hit [Enter]; otherwise, backspace the suggestion away.*

>> *win_win = 3*

>> *2win = 3*

>> *win-win = 3*

>> *win.win = 3*

Let's review those last three entries:

Invalid variable name	Result	Reason
2win	Error: Unexpected MATLAB expression.	Variable begins with a number.
win-win	Error: The expression to the left of the equals sign is not a valid target for an assignment.	MATLAB thinks you're trying to subtract a variable called win from itself.
win.win	A new structure called *win* is created with a member called *win*.	The variable name contains an invalid character, but it is a valid way to refer to a member of a structure (a more advanced data type that we'll talk about in a later tutorial). MATLAB thinks you meant that and creates it.

Effective Variable Names

Strive to write code that is easily understood by another programmer. Right now, this probably isn't an issue because you're doing your work solo, but in the real world most programs are group efforts. Other programmers will review your code before adding it to a larger project, and later programmers will look at it to maintain and upgrade it over time, possibly when you're no longer available to consult. One way to make your code understandable is to use descriptive variable names. In our examples so far, we've stuck to simple one-letter variables that don't mean anything. Later, we'll use variable names that explain the variable's purpose and/or units.

For instance, say that you're working on a simulation for a Mars lander and need to know the distance from the surface to the lander. You might name your variable *altitude*, and that's not bad (better than *x* anyway), but if altitude = 4500, the question "4500 whats?" comes to mind. An American programmer might assume miles, while a Bulgarian programmer might assume kilometers. At least one of those two is going to see a spectacular crash. Perhaps *altitude_in_meters* is a better name, or *altitudeInMeters* if that's your preferred style.

At the same time, don't get carried away. You're going to have to type that variable name over and over, so you'll quickly learn to keep them succinct.

Tip: *You can drag and drop variable names from the Workspace into the Command Window (and Editor also) to save having to type them.*

Storing Numeric Values

All variables have a *class* or type, which is the sort of data it holds: numbers, strings, structures, etc. In this section, we'll talk about variables that hold numbers. If you remember from math class, integers are whole numbers (4, 288, 42), while floating-point numbers (also called "real" numbers) contain decimal values (3.1417, 6.125, 9.9).

Type the following into the Command Window:

> >> x = 666
>
> >> class(x)

Hmm. Even though you clearly entered an integer value, MATLAB created a variable of type *double*. Unless you explicitly specify the type of your numeric data, this is MATLAB's default behavior.

Back to the Command Window, enter these lines:

> >> x = int8(666);
>
> >> class(x)

Here, you specifically told MATLAB that you wanted to store the number 666 as an 8-bit integer called *x*. MATLAB has a function, int8, that converts the number (or variable) you provide as an *argument* (the value in parentheses) to an 8-bit integer. The class function reports back the class or type.

The number of bits is how many binary digits the value stores. For example, 8 bits can store a binary number such as 11011011. If the sign (positive or negative) takes up one binary digit, then the value that can be stored is reduced.

2^7	2^6	2^5	2^4	2^3	2^2	2^1	2^0	
1	1	1	1	1	1	1	1	
128 +	64 +	32 +	16 +	8+	4 +	2 +	1+	= 255

There are several *integer types* for different levels of size and precision. Each comes in a *signed* and *unsigned* version. If you're only dealing with positive integers, using an unsigned type allows you to specify numbers twice as large.

Integer Types			
Type	Bits	Signed or Unsigned?	Range of values
int8	8	signed	-128 to 127
uint8	8	unsigned	0 to 255
int16	16	signed	-32,768 to 32,767
uint16	16	unsigned	0 to 65,535
int32	32	signed	-2,147,483,648 to 2,147,483,647 (Approx. -2.1 to 2.1 billion)
uint32	32	unsigned	0 to 4,294,967,295
int64	64	signed	-9,223,372,036,854,775,808 to 9,223,372,036,854,775,807 (Approx. -9.2 to 9.2 quintillion)
uint64	64	unsigned	0 to 18,446,744,073,709,551,615

Tip: *The* class *(variableName) command is very useful, particularly in troubleshooting.*

Tip: *Notice the difference between int8( ) and uint8( ). The unsigned uint can store a number twice as large because it doesn't have to store the sign.*

Tip: *Additional handy functions are intmin( ) and intmax( ). Supply a type to one of those functions and MATLAB will tell you the minimum or maximum value that type can hold. For instance, intmax('int32').*

Floating-point types can represent numbers so big that they make an int64 look like something you could count on your fingers. This is possible because part of their stored value is an exponent. They come in two varieties.: single and double.

Floating Point Types			
Type	Bits	Signed or Unsigned?	Range of values
single	32	signed	-1.79×10^{38} to 1.79×10^{38}
double	64	signed	-1.79×10^{308} to 1.79×10^{308}

Constants

> **Tip:** *You can define a class and give it the* constant *property, but if you are contemplating that, you are way ahead of this introductory book.*

A *constant* is a value that, once assigned, cannot be changed later. MATLAB doesn't have a straightforward way for you to declare your own constants, but it does provide a handful of its own.

There are two special floating-point values to be aware of. You'll see these values as the result of an operation. Type the following statements into the Command Window and watch the result:

> *>> 3/0*

> *>> inf-inf*

> **Tip:** *Inf and NaN are not case-sensitive. Inf or inf, as you please. Imaginary numbers also have a special constant, i, used to represent the imaginary portion of the value.*

In many programming languages, an error results if you attempt to divide by zero, but not MATLAB. It returns the constant value *Inf* for infinity, or *-Inf* for negative infinity. *Inf* can be the result of any operation that results in enormously large or small numbers.

If you've tried to perform an operation that is not mathematically defined, you'll see the constant value *NaN*. This means "not a number." This isn't quite a constant by any standard definition, but it isn't quite anything else either. In some matrix cases it's used as a placeholder.

We'll show you how to make use of these values when we discuss *if* statements in a later tutorial. The other constant of note is *pi*. You've already used it in an earlier tutorial. Try this in the Command Window:

> *>> format long*

> *>> pi*

> *>> pi + 7*

Remember, format long tells the Command Window to display numbers with more digits. This is for display purposes only. In calculations, the full stored number is used.

Exceeding a Type's Range

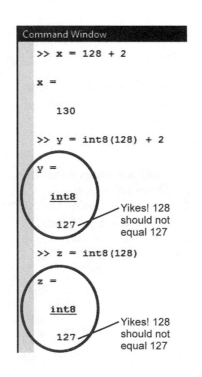

Now for a chilling demonstration of where a type can cause things to go terribly wrong. Enter the following into the Command Window:

> *>> x = 128 + 2*

> *>> y = int8(128) + 2*

> *>> z = int8(128)*

Well, that's not good! Notice in the Workspace that z has a value of 127, not the 128 you'd specified. If you provide a number larger or smaller than the type can handle, MATLAB returns the largest or smallest number that type can provide, with no error. 127 is as high as an int8 can go. This can lead to miscalculations that can be hard to track down. Be aware of the magnitude of your data versus your data types.

You're probably wondering why we don't just use doubles all the time. You probably can – no harm, no foul. However, if you're working with Big Data, you may work with data sets that contain billions of numbers or more. Code that is optimized for its expected data can prevent memory overruns and significantly affect run time. It literally can be the difference between getting results in minutes and getting them in years.

Numerical Functions

As you can imagine, there are almost more built-in MATLAB functions for manipulating numbers than there are stars in the sky (oh, perhaps a bit of hyperbole there). Here are some functions you might find useful.

Function	Action
ceil	Rounds toward positive infinity
floor	Rounds toward negative infinity
fix	Rounds toward zero
round	Rounds toward the nearest whole number
mod (a, b)	Returns the modulus of *a* divided by *b*. Retains the sign of the divisor. Example: mod(10, -7) returns -4.
rem (a, b)	Returns the remainder in a division operation. Retains the sign of the dividend. Example: rem(10, -7) returns 3.

Type the following commands into the Command Window and note the output that results.

Command Window Entry	Result
>> ceil(3.4)	ans = 4
>> ceil(-3.4)	ans = -3
>> floor(3.4)	ans = 3
>> floor(-3.4)	ans = -4
>> fix(3.4)	ans = 3
>> fix(-3.4)	ans = -3
>> round(3.4)	ans = 3
>> round(-3.4)	ans = -3
>> mod(5, 2)	ans = 1
>> rem(5, 2)	ans = 1
>> mod(5, -2)	ans = -1
>> rem(5, -2)	ans = 1
>> mod(-5, 2)	ans = 1
>> rem(-5, 2)	ans = -1

Tip: *Functions generally have some input or "argument" that they act on. For example:*
ceil (x)
where x is the "argument" or input that can be provided to the function. You provide the input argument inside parentheses after the function name. Use MATLAB help to look up these details when you use a function. When we write ceil() it is to remind you that you will provide an input when using it. Some functions may accept multiple arguments, which may also be left out depending on the use.

Tip: *Don't forget the parentheses! Without them the value 3.4 is assumed to be three text characters: a 3, a period, and a 4.*
>> ceil(3.4)
    ans = 4
>> class (3.4)
    ans = 'double'

>> ceil 3.4
    ans = 51   46   52
>> class 3.4
    ans = 'char'

There are also conversion functions for every numeric type. For example int16(43) converts 43 (a double by default) into a 16-bit integer. One word of warning with these functions: beware of losing precision and exceeding type ranges in the conversion, particularly when converting unsigned types to signed types.

Enter the following commands in the Command Window

>> *x = uint8(255)*

>> *y = int8(x)*

An int8 can't hold the number 255, so the number becomes 127. Probably not what you wanted!

Of course, standard *trigonometric functions* are available as well, each in two versions: one which returns the value in degrees and one which returns the value in radians.

*On your own, browse the MATLAB help for **Elementary Math** (Figure 2.2) then select **Trigonometry** from the results to see the full list of trig functions. Explore the other functions, too.*

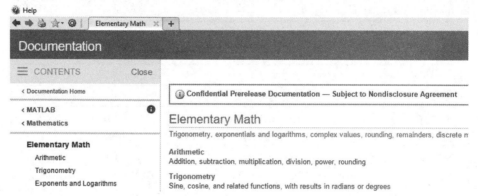

Figure 2.2

Strings

A *string* type holds text. An example of a string is this very sentence. A string can hold any valid character. "123456" can be a string. "< <= > >=" is another. To discuss strings, we need to talk about *arrays*. We'll go into arrays and matrices in more depth in a later tutorial, so here we'll just cover the concepts you need right now.

Array Basics

An *array* is a collection of data, all of the same type. Each item in the array has an *index* by which it can be accessed. Consider the phrase "Carpe dium." Each letter, space, and punctuation mark in that phrase is a character. If you store that string as an array of characters, this is how it is stored:

Index	1	2	3	4	5	6	7	8	9	10	11
Character	C	a	r	p	e		d	i	u	m	.

This a one-dimensional array of 11 characters (1 x 11).

Type the following lines into the Command Window:

>> st = 'Carpe dium.'

>> size(st)

>> length(st)

>> st(4)

>> st(4) = '9'

>> st(4) = 33

The function size() tells you that it is a 1 x 11 array, while the function length() simply tells you that the array contains 11 elements. You accessed a specific element in the array with st(4) and overwrote that character with st(4) = '9'.

The statement st(4) = 33 had the surprising result of overwriting that element with an exclamation point. This is because all elements in an array must be of the same type – characters, strings, numbers, etc. The variable st is an array of characters, so you can't overwrite an element with a numerical value. Instead, MATLAB assumed that you wanted *ASCII character* 33, which is an exclamation point. The American Standard Code for Information Interchange (ASCII) set is a collection of characters, each with its own numerical code, commonly used in computing to represent letters and symbols (including numbers). The standard ASCII set includes all the characters you can type on a US keyboard, each encoded in 7-bits. Extended ASCII sets exist for characters such as accented, Asian, Arabic, Cyrillic, and Hebrew letters, scientific and other specialized symbols, and many more. The following table shows some of the ASCII character set. You can find the entire sequence on the web.

```
Command Window
>> st = 'Carpe dium.'
st =
    'Carpe dium.'
>> size(st)
ans =
     1     11
>> length(st)
ans =
    11
>> st(4)
ans =
    'p'
>> st(4) = 33
st =
    'Car!e dium.'
>> st(4) = 33
st =
    'Car!e dium.'
```

Tip: *The size() function returns information as rows first, then columns.*

Decimal	7-bit Binary	ASCII Character (notes)
0	0000000	NUL '\0'
1	0000001	SOH (start of heading)
2	0000010	STX (start of text)
3	0000011	ETX (end of text)
4	0000100	EOT (end of transmission)
10	0001010	LF '\n' (new line)
32	0100000	SPACE

Decimal	ASCII							
33	!	57	9	68	D	120	x
34	"	58	:	90	Z	121	y
35	#	59	;	91	[	122	z
36	$	60	<	92	\ '\\'	123	{
37	%	61	=	93	]	124	|
38	&	62	>	94	^	125	}
39	'	63	?	95	_	126	~
48	0	64	@	96	`	127	DEL
49	1	65	A	97	a		
50	2	66	B	98	b		
		67	C	99	c		

The Two String Types

There are two types of strings: a *character array* and a *string array*. Strings as character arrays are entered by surrounding the text in single quotes ('the'). Strings as string arrays are entered by surrounding the text in double quotes ("the"), resulting in an array with one element.

String arrays are new to MATLAB 2017. They are more memory-efficient and have a number of new functions that can be used with them, as well as using many of the familiar character array functions. Which you choose in any situation will depend on your programming needs.

Enter the following lines in the Command Window for a whirlwind tour of some differences between character arrays and string arrays:

```
>> clc
>> sc = 'This is a character array'
>> st = "This is a string array"
>> class(sc)
>> class(st)
>> size(sc)
>> length(sc)
>> size(st)
>> length(st)
>> sc(1)
>> st(1)
>> sc(2) = 'H';
>> sc
>> st(2) = "And this is the second string in the array";
>> st
```

The major difference to remember is that a string as character array is a 1 x *length* array of individual characters, while a string as a string array is a 1 x 1 array that contains a single string.

String and Character Functions

MATLAB provides many functions to use with strings and characters. Here's a list of some of the most useful. For a complete list and usage particulars, search for "Characters and Strings" in the MATLAB help.

Function	Use with...	Action	Return type
ischar	any variable type	Is this a character array?	logical
isstring	any variable type	Is this a string array?	logical
isletter	character arrays	Which characters are letters?	logical array

isspace	character arrays	Which characters are spaces?	logical array
length	character arrays	How long is the array?	integer
strlength	character or string arrays	How long is each string in the array?	integer array
lower	character or string arrays	Convert all letters in the string to lower-case	string
upper	character or string arrays	Convert all letters in the string to upper-case	string
strcmp	character or string arrays	Compare two strings for equality. Use this instead of =.	logical
strcat	character or string arrays	Add one string to the end of another	string
startsWith	character or string arrays	Does the string start with a particular substring?	logical
endsWith	character or string arrays	Does the string end with a particular substring?	logical
contains	character or string arrays	Does the string contain a particular substring?	logical
strfind	character or string arrays	What is the starting position of a substring?	integer
strtrim	character or string arrays	Remove any spaces from the beginning and end of the string	string
replace	character or string arrays	Replace a substring with a new substring	string
erase	character or string arrays	Remove a substring from the string	string
insertBefore	character or string arrays	Add a new substring before a specific point in the string	string
insertAfter	character or string arrays	Add a new substring after a specific point in the string	string
split	character or string arrays	Split the string into two or more substrings	strings

Type the following statements into the Command Window, noting the form of the functions and the values they return:

>> **charArray = 'This is a character array.'**

>> **stringArray = "This is a string array."**

>> **ischar(charArray)**

>> **ischar(stringArray)**

>> **isstring(stringArray)**

The results you see in the Command Window are just as you'd expect.

Tip: *Use the MATLAB help to look up the necessary format for these functions when necessary. For example, the syntax for the insertBefore is: newStr = insertBefore(str,endStr,newText). In this case, str, endStr, newText are the inputs for the function.*

Syntax *is like the exact recipe you must follow for MATLAB to understand the command input. We assume you can look these up in the help when you need them. What kind of person memorizes every instance of every function anyway?*

Tip: *Several of the string functions have a separate version that ignores case. For instance, strcmp (case-sensitive) and strcmpi (case-insensitive).*

Tip: *Keep in mind that functions and variable names are case sensitive. The name "charArray" is not the same as "Chararray".*

Tip: *In functions such as contains and strfind that take a substring as an argument, you can use either a character array (single quotes) or a string array (double quotes). MATLAB knows what you mean and takes care of any necessary conversions behind the scenes.*

Tip: *Use clc to clear the Command Window on your own when things get a bit too messy.*

```
Command Window
>> x = 3
x =
     3
>> class (x)
ans =
    'double'
>> x = 'I am a string'
x =
    'I am a string'
>> class (x)
ans =
    'char'
```

Tip: *Use clear when you want to empty the variables from the Workspace. This will delete the values if they are not saved to a file, so make sure this is what you want before clearing the Workspace. You do not need to use clear to set the variable to a different value, but this also means that you can accidentally overwrite a value when you reuse a variable name.*

Now let's get into trickier stuff. Type the following line into the Command Window:

> >> *isletter(charArray)*

Interesting. This function returns a numerical array of the same size as the character array. In each position is a logical value telling you whether that character is a letter (a-z, A-Z), 1 for true, 0 for false.

> **Give** *isspace* **a try on your own.**

Enter this command into the Command Window:

> >> *upper(stringArray)*

Notice that it shouted "THIS IS A STRING ARRAY" back at you in the Command Window, but the variable *stringArray* in the Workspace hasn't changed. This is because we didn't assign the returned value back to a variable. Try this instead:

> >> *stringArray = upper(stringArray)*

Lesson: Just entering the function does not store the value for later use. If you want to keep the results, use the equals sign and store the result in a variable (which you can think of as a named storage location)

Now let's talk about what's arguably the most important string function: strcmp. Type the following into the Command Window:

> >> *text1 = 'Four score and seven years ago'*

> >> *text2 = '87 years ago'*

> >> *text3 = 'Four score and seven years ago'*

> >> *strcmp(text1, text2)*

> >> *strcmp(text1, text3)*

> >> *text1 == text3*

Yipes. And that's why you need strcmp. The relational equality operator (==) looks at the string as an array, compares the individual characters, and returns an array of logical results for each index. The strcmp function answers the question you're really asking – "Do these two variables contain the same text?"

The situation is a little cloudier with string arrays, but the general advice remains the same: whenever you want to compare two strings, use strcmp.

Enter the following commands into the Command Window:

> >> *contains(text1, 'seven')*

> >> *strfind(text1, 'seven')*

The first command tells you that yes, the substring 'seven' is in text1 somewhere. The second command tells you where it is – the character 's' of 'seven' is at index 16.

Finally, type the following line into the Command Window:

> >> *insertBefore(text1, strfind(text1, 'seven'), 'fifty-')*

This command demonstrates two important points. First, a function can be used like a value within another function. The command is equivalent to the following two commands. We'll discuss the pros and cons of both forms when we talk about functions in a later tutorial.

```
index = strfind(text1, 'seven')

insertBefore(text1, index, 'fifty-')
```

The second point the original command made was about nested parentheses. This can occur in all kinds of situations and is another common source of simple coding errors. MATLAB helps you to avoid this by highlighting the matching opening parenthesis when you type a closing parenthesis. Also, if you wind up with a mismatched number of opening and closing parentheses, the error message that results will usually point you to that as the issue.

MATLAB and Type Flexibility

MATLAB is not a strongly typed language. That means that you can perform assignments and operations on variables of wildly different types, usually without causing an error. This can be a blessing in that you can usually avoid explicitly defining the type for each variable, but it can be a curse when the result of, say, adding a number to a string, is puzzling (though it can all be explained).

Type the following commands, and note the results.

```
>> x = 3
>> class(x)
>> x = 'I am some text'
>> class(x)
```

The variable becomes the type of whatever you assign to it, regardless of what it held before. Not many languages allow that.

Now enter these commands in the Command Window. Use the up arrow on your keyboard and select from the history to save yourself some typing.

```
>> x = 3
>> class(x)
>> y = single(5)
>> z = x + y
>> class(z)
>> x = int16(5) + int8(3)
```

The last line demonstrates that this doesn't work with integer types, only floating point types.

```
Command Window
>> x = 3
x =
     3
>> class (x)
ans =
     'double'
>> y = single (5)
y =
   single
      5
>> z = x + y
z =
   single
      8
>> class (z)
ans =
     'single'
>> x = int16(5) = int8(3)
 x = int16(5) = int8(3)
                  ↑
Error: The expression to the left of the equals sign is not a valid target for an
assignment.
```

Figure 2.3　　　　**Error Resulting from Conflict of Integer Types**

Finally, enter the following into the Command Window:

>> *x = 3*

>> *y = 'I am a string'*

>> *z = x + y*

>> *class(z)*

>> *char(z)*

```
>> z = x + y
z =
      76      35     100     112      35     100      35     118     119
>> class(z)
ans =
     'double'
>> char(z)
ans =
     'L#dp#d#vwulqj'
```

Figure 2.4　　　　**Adding Numbers and Letters**

MATLAB looked at the string as a character array, added 3 to each character's ASCII value, and returned a numerical array of those values as numbers instead of the characters. The final line, char (z), casts the array back to a character array-type string based on the ASCII values. Potential uses in encoding, perhaps – a simple alphabet shift. MATLAB allows a lot of flexibility, and you can use this in creative ways to accomplish interesting solutions. Just make sure you're doing it on purpose! Use the functions available to set the data type to prevent problems.

Now that you have been introduced to types and strings, let's take a look at an example of importing data into MATLAB and some options for data types.

Genetic Data Example

Humans typically have 23 pairs of chromosomes located in each of their cells. One of the shorter ones, chromosome 22, has more than 50 million base pair building blocks. Base pair data is encoded by one of four letters: G, C, A, and T (guanine, cytosine, adenine, and thymine). A variation in the expected code, or a missing code, may result in health problems. Customizing medicine for an individual's specific genetic variation is a hot topic in improved therapies.

The data file, *Chrome22- HG00096.vcf*, included in our download, is a Variant Call Format (vcf) file. It contains information about positions in the genome and genotype information for each sample. It is a text file that starts with a header section followed by this required data on each line:

CHROM: chromosome number. The data file you will use only has chromosome 22 data.

POS: position. The number that represents the location of the gene on the chromosome. If you look at the .vcf file you will see that the range is 1 up to about 1.2 million. (This data should be a number with no decimal portion.)

ID: identifier. The dbSNP variant is given as an rs number(s). No identifier should be present in more than one data record. If there is no identifier available, then the value '.' should be used. (This should be a string with no white-space or semi-colons permitted.)

REF: reference base(s). Each base allele must be one of A,C,G,T, or N (meaning aNy of these.) This data is not case-sensitive. More than one base letter is allowable. The value in the POS field gives the position of the first base in the string. Gene variations may have insertions or deletions in which either the REF or the ALT alleles are null/empty. POS denotes the coordinate of the base preceding the polymorphism (a big word meaning there can be several different forms, one of the things you would be searching for in genetic variation). (This required data should be a string.)

ALT: alternate base(s). Similar to REF, but for alternate non-reference alleles. These alleles do not have to be called in any of the samples. Options are A,C,G,T,N, or *. The '*' is used to indicate that the allele is missing due to an upstream deletion.

QUAL: quality. Quality score based on the Phred-scale. Our data all has 100 for the quality. You can read more about this topic on your own or in a statistics class.

FILTER: filter status. "PASS" indicates the quality score is passing.

INFO: additional information. Keys such as AA (ancestral allele), AC (allele count in genotypes), and AF (allele frequency) are often used, but others are permitted.

Any other columns in the data file are optional. So now you know a bit about .vcf files, so let's get to importing one into MATLAB.

Importing Data into MATLAB

MATLAB has a handy feature for importing data using the Import Data tool. You can also use readtable, csvread, dlmread, textscan, imread, and other command entries. For this example, we will use the Import Data tool from the Home tab of the ribbon.

Tip: *Don't worry, nobody is going to test you on your knowledge of genetics or vcf files here. We are just going to import some data from a large file and check out some things about data types. It's fun to look at real data and who knows, maybe you will go find a cure for cancer or schizophrenia in your spare time.*
You can read more about rs numbers in genetics at en.wikipedia.org/wiki/DbSNP and many other places.
The variation viewer at www.ncbi.nlm.nih.gov/variation/view is a great tool for exploring genetic information.

Tip: *Can't remember which letters are in the genetic code? Make up a mnemonic, like GAG-A-CAT. Its all Gs, As, Cs and Ts. The N is for aNy.*

Import Data

Click: **Import Data** *from the ribbon Home tab*

Figure 2.5   **Ribbon Home Tab with Import Data Tool**

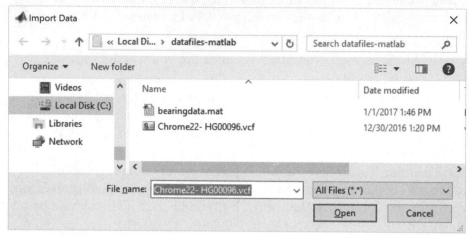

Figure 2.6   **File Browser**

WARNING: *The file you will import from has more than 1 million rows in it. If you are not working on a fast computer system, use the smaller data file instead.*

Show all files

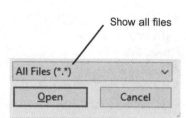

On your own, change the file type to All Files (*.*) at the bottom right of the browser window, to show the .vcf file.

Use the file browser to locate and open the data file **Chrome22-HG00096.vcf** *OR use the smaller data file named* **Chrome22-small.vcf** *if your system will not handle larger files easily.*

The import data window opens on your screen similar to Figure 2.7. It shows a view of the data similar to spreadsheets you may have used.

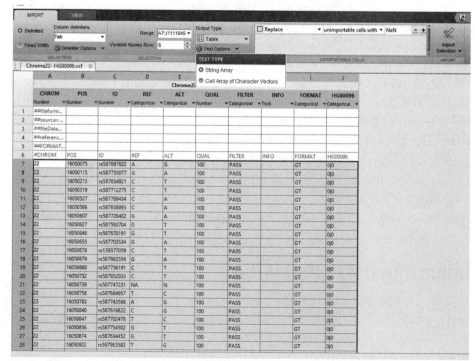

Figure 2.7   **Data to Be Imported**

Text Options

Notice under the drop down for Text Options, there are options to create a String Array or a Cell Array of Character Vectors, similar to entering strings using either double-quotes ("string array") or single-quotes ('character array').

Leave the Text Type set to String Array.

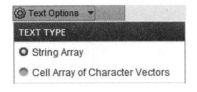

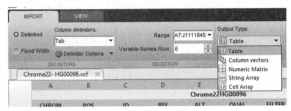

Figure 2.8   **Output File Options**

Output Type

Our data will work best as a table which has rows and columns, but column vectors, numeric matrix, and our old friend the string array and cell array are available options for data that would be suited to those.

*Leave **Table** set as the **Output Type**.*

Delimiters

Notice there are also options that allow you to select whether your data has delimiters, such as commas or tabs, between the entries. Our data does, so leave this set to Tab delimited. The other option is to have a fixed width of characters for each entry. This is useful when each entry is always the same length, like a list of numbers or when the data itself contains commas or tabs and you don't want the columns based on those.

Variable Names Row

This option lets you select a heading row to generate the variable names used for the columns of the table. We will use row 6, which contains headings CHROM, POS, ID, ALT, QUAL, FILTER, etc.

*On your own set **Variable Names Row** to row 6.*

Range

You don't have to use all of the data in your giant file. The COMT gene is associated allele variants such as rs4680, rs737865, and rs165599 which may be associated with response to antidepressants among other things.

The COMT gene is located starting at 19,941,740 up to 19,969,975, so we will extract only a section of the data around this gene to keep the imported file smaller, just for this project. We've already figured out which rows are near this region to make this easier.

*On your own, set the **Range** to **A106710:J107844**.*

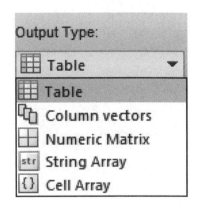

When the file is imported, only those rows (106710 to 107844) and columns (A-J) will be used in the MATLAB table.

Import Selection Options

Now you are ready to import the data into MATLAB, where you will see it as a new variable in the Workspace. Notice that the Import Selection button has a small triangle at the bottom that you can click to expand to see more options.

We will use Import Data, but you can also use this to generate a script, which you used in an earlier tutorial, or a function, which you will learn about in a later tutorial. This is very handy if you must convert multiple data files. We are just doing this once, so...

*Click: **Import Selection***

The data is imported to the MATLAB table and appears in the Workspace as shown in Figure 2.9.

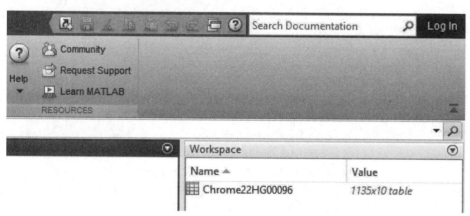

Figure 2.9 **Table Imported in Workspace**

*Double-click: **Chrome22HG00096** in the Workspace list*

The table of imported data opens and the Command Window shows below it. Now you can enter MATLAB commands to interact with the data.

*On your own, **click and drag on the divider to resize the Command Window larger.***

Enter the following command at the prompt to sort the data by 'ID', the variable heading for row 3.

*>> **tblIDs = sortrows(Chrome22HG00096,'ID')***

The table information displays in the Command Window, sorted by the ID column values. Notice the new variable, tblIDs, in the Workspace.

Tip: *If you used the short file, then the variable names will be in row 1.*

Tip: *If you used the short file, then select the entire range of the data.*

Tip: *The Import Data tool generates a script that you can edit and run to import files. This is a handy way to import multiple files with similar data quickly.*

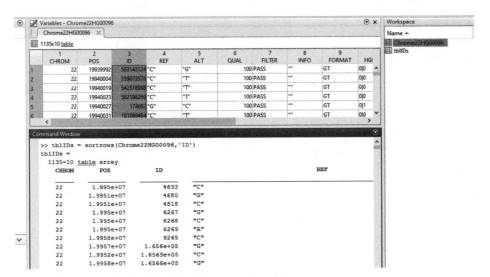

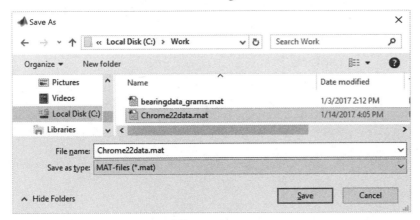

Figure 2.10   Sorted Data in Command Window

Remember that unless you save the imported file will not be saved when you either clear your variable list or exit MATLAB. Here's another way to save a file.

*Right-click: **Chrome22HG00096** in the Workspace.*

*Use the context menu to select **Save As...***

Browse to your Work folder and save the file with the name Chrome22data.mat as shown in Figure 2.11.

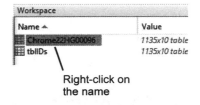

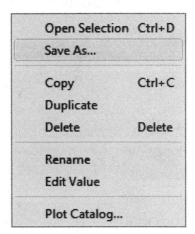

Figure 2.11   Saving the File

Congrats! You have finished Tutorial 2.

Tip: Check out the interface and notice some other ways you could sort this table.

Right-click on the name

Key Terms

argument	*equal sign operators*	*operator precedence*
arithmetic operators	*floating-point types*	*pi*
array	*index*	*relational operators*
ascii character	*Inf*	*signed*
assigment statement	*-Inf*	*string*
character array	*integer types*	*string array*
class	*logical operators*	*trigonometric functions*
constant	*logical type*	*unsigned*
double	*NaN*	*variable*

Key Commands

&&	format	length	startsWith
|	imread	logical	strcat
and	insertAfter	lower	strcmp
ceil	insertBefore	mod	strfind
class	int16	not	strlength
contains	int8	or	strtrim
csvread	ischar	readtable	textscan
dlmread	isletter	rem	upper
endsWith	islogical	replace	xor
erase	isspace	round	
fix	isstring	size	
floor	length	split	

Exercises

Exercise 2.1

Based on operator precedence, predict the result of the calculation below without using Matlab. Once you think you know the answer, enter the statement in the Command Window. Do your answers agree?

$$5 + 2 * 8 / 2 - (3 * 2 + 10) / -1$$

Exercise 2.2

Predict the value of z after the following statements:

x = 1

y = 5

z = ~(x < y || ~(y < x) && islogical(x))

Enter the commands above into the Command Window. Was your prediction correct? You have a 50:50 chance of getting it right, so make sure that if your prediction was correct, it was correct for the right reasons.

Exercise 2.3

Predict both the value and the data type of x in this equation:

x = 55 + uint32(-22) + pi

Enter the command in the Command Window. Were you right? What does this equation demonstrate about variable types?

Exercise 2.4

Predict the value of each of these commands:

int8(ceil(127.1))

int8(floor(127.9))

int8(fix(127.5))

int8(round(127.7))

int8(ceil(rem(-528.6,200)))

Enter the commands in the Command Window and compare each result to your prediction. If you were mistaken on any, enter them in parts from the center outward, such as ceil(127.1) then int8(ceil(127.1)), to see where the number did something you didn't expect.

Exercise 2.5

Predict what you expect to see (generally, not precisely) as a result of the last two lines of this set of commands:

x = 'small kittens'

y = "small kittens"

3 + x

3 + y

PROGRAMMING BASICS: ARRAYS AND STRUCTURES

3

Introduction

The data types we've talked about so far – integers, floating-point numbers, and strings – each conceptually holds a single number or string. Now, welcome to the world of *complex data types*! A complex data type contains a collection of data. It might be a gang of integers, a plethora of strings, or even a mixed bag of any and all data types.

Arrays

You had a brief introduction to arrays in the last chapter, but, given the importance of arrays in MATLAB, that was like meeting the king at a party and thinking he was a waiter. To review, an array is a collection of a single type of data – characters, integers, and so on. In MATLAB, you can even have an array of arrays. Each value in the array is called an *element*. Each element can be accessed through an array *index*, a number or series of comma-separated numbers in parentheses that indicate row, column, page, and beyond, such as exampleArray(4,5).

There are some special array types that have their own names and are progressive subcategories:

A *matrix* is a two-dimensional array, indexed in rows and columns.

A *vector* is a 1 x *n* (*n* being any number) or *n* x 1 matrix – one row or one column. A *row vector* is 1 x *n* and a *column vector* is *n* x 1. For instance, strings as character arrays are a vector.

A *scalar* is a 1 x 1 vector, such as a single character or number. An int32 variable is a scalar.

Calling the number 57 a scalar, a vector, a matrix, or an array would all be technically correct, and there are functions to prove it.

Start MATLAB if you haven't done so and enter these commands in the Command Window:

> >> *isscalar(57)*
>
> >> *isvector(57)*
>
> >> *ismatrix(57)*

All return *true (1)*. There's no *isarray* function, but because scalar, vector, and matrix are all array types, as far as MATLAB is concerned 57 is an array well.

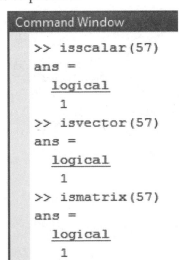

```
Command Window
>> isscalar(57)
ans =
  logical
   1
>> isvector(57)
ans =
  logical
   1
>> ismatrix(57)
ans =
  logical
   1
```

Figure 3.1   Command Results

Objectives

When you have completed this tutorial, you will be able to

1. Understand what constitutes an array.

2. Access array elements using indexes.

3. Explain the difference between an array, matrix, vector, and scalar.

4. Define a series of numbers using the colon (:) operator.

5. Access selected array elements in multiple dimensions using the colon (:) operator and the end keyword.

6. Explain the difference between an array and a cell array.

7. Use the MATLAB Variables window to examine and modify array and structure elements.

8. Create a structure and add and modify its fields.

9. Work with an array of structures.

Note: *Arrays can have as many dimensions as needed, though after three or four it gets hard for most human brains to conceptualize them. Still, they have their place, and it's called physics.*

```
>> A = randi(10,2)
A =
     10      9
      5      2
>> A(4,3) = 54
A =
     10      9      0
      5      2      0
      0      0      0
      0      0     54
```

Figure 3.6

```
>> A(3,4) = 16
A =
      0      0      0      0
      0      0      0      0
      0      0      0     16
```

Figure 3.7

```
>> A = []
A =
     []
>> size(A)
ans =
      0      0
>> isempty(A)
ans =
  logical
    1
```

Figure 3.8

One outcome of this hierarchy is that you can send any array type to any function that deals with its specific type or one of its parent types. For example, you can send 57 to any function that deals with scalars, vectors, matrices, or arrays. Going forward, we'll usually use the more specific terms, but we'll also use "array" generically when a point applies to any array type.

Growing Arrays

To increase the size of an existing array, you can simply assign a value to an array element beyond its current array size. Because an array must be rectangular, all of the other elements necessary to meet that requirement are also added, with a value of 0.

The randi function generates pseudorandom integers. When you use the format X = randi(imax,n), it returns an n-by-n matrix of integers on the interval from 1 to *imax*. We will use it in the next step to add some values into a matrix we will name *A*.

Enter the following into the Command Window:

>> *A = randi(10,2)*

>> *A(4,3) = 54*

Your generated numbers may be different, but the result of assigning a value to the element in the fourth row, third column of a 2x2 matrix is that the array expands as needed, filling the unassigned elements with zeros similar to Figure 3.6.

Similar to growing an existing array, assigning a value to an element in a non-existent array creates that array with the new value in the lower right corner. Try this:

>> *clear*

>> *A(3,4) = 16*

The Empty Array

Enter these commands in the Command Window:

>> *A = []*

>> *size(A)*

This is an empty array: an array with no elements. The empty array is worth mention because you might see it as an operation's return value, which indicates that the operation has no answer. You can also use it to initialilze an array. You can test for an empty array by using the isempty function. Try it now.

>> *isempty(A)*

Matrix Basics

Keep this in mind:

To MATLAB, almost everything is a matrix

To reiterate, a matrix is a two-dimensional grid of values, indexed in rows and columns. Here's an old friend:

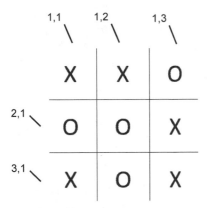

Figure 3.2   **Tic-Tac-Toe**

The tic-tac-toe game "board" starts as an empty 3 x 3 matrix: three rows and three columns. This matrix is filled with one character (an X or an O in this case) in each element. Each element is referred to by its (row, column) coordinates, beginning with the upper left corner as (1,1).

Let's get back to MATLAB seeing everything as a matrix. Type the following into the Command Window:

```
>> aScalar = int16(345);
>> string1 = 'To be or not to be';
>> string2 = "that is the question";
>> aVector = 1:5;
>> class(aScalar)
>> class(string1)
>> class(string2)
>> class(aVector)
```

No surprises so far: an integer, a character array, a string array, and a vector array of double-precision floating-point numbers. Now type the following into the Command Window:

```
>> size(aScalar)
>> size(string1)
>> size(string2)
>> size(aVector)
```

The **size** function returns array dimensions, row first, then columns. The variable *aScalar*, a single number of type int16, is considered by

Note: *In MATLAB, array indexes are 1-based, which means that the first item is (1), (1,1), (1,1,1), and so on. Some programming languages such as C++ and Java are 0-based, so the upper left corner of a matrix in those languages would be (0,0). For example, in a 1-based language, the number of columns is x and the index of the last column is also x. In a 0-based language, the number of columns is x and the index of the last column is x-1.*

MATLAB as a matrix with one row and one column (ans = 1 1), which is to say a scalar. The vector aVector (ans = 1 5) is a matrix of one row and five columns, each of which contains a single number.

Strings as Matrices

The difference in the matrix size of the two strings above is important to your understanding of those types. The character array string (using single quotes, ans = 1 18) is a matrix of one row and eighteen columns, each of which contains a single letter. The string array (using double quotes, ans = 1 1) is a matrix of one row and one column: a single element that contains the entire string. Enter these commands in the Command Window:

> > *string2 == 'b'*

> > *strcmp(string2, 'b')*

> > *string1 == 'b'*

> > *strcmp(string1, 'b')*

The == operator and the strcmp() function give the same result for string array *string2*. They give very different results for the character array *string1*. Because MATLAB sees *string1* as a 1 x 18 matrix, it makes the equality comparison on each individual element, returning a 1 x 18 matrix of those logical results. The character at *string1(1,4)* is a 'b', so element (1,4) in the result matrix contains *true* (1). The strcmp() function, on the other hand, asks whether the two strings are the same and gives you the single answer *false* (0).

One effect of this is that you can use the == operator with a string array and get the right answer. However, consistent use of the strcmp() function regardless of the string type ensures that you'll never accidentally use == on a character array when you didn't mean to. This also shows that using the == operator on a character array can return useful results, if those are the results you're looking for.

The (:) Colon Operator

You were briefly introduced to the colon operator (:) when you defined an array in the last chapter. The colon operator is used to specify an evenly spaced series of numbers in a conveniently terse manner. The numbers specified can be used in many ways, from specifying the contents of a new array to indexing elements in an existing array to providing a counter for a loop. Its general form is this:

first number : step amount : limit

The *first number* is exactly that, the first number in the series.

The *step amount* is the numerical distance between each number in the series. For instance, if the first number is 1 and the step amount is 2, the series will be 1, 3, 5, 7, 9, and so on. You can also have a negative step amount if *limit* >= *first number*. **The step amount is optional**. If you leave it out, for instance 5:10, the default step amount is 1.

Tip: *Don't forget that a space, or anything else that isn't a letter or a number, counts as a character in a string. In the string 'To be or not to be', the element (1,3) is a space.*

The *limit* is the maximum number that can be found in the series (or the minimum if the step amount is negative). The step amount can cause this number to be exceeded but not hit precisely, so it might not be included in the resulting series. To see what we mean try these next statements. Enter the following in the Command Window:

```
>> 1:5
>> 1:2:5
>> 1:2:6
```

The specification 1:2:6 produced the series 1, 3, 5, 7, but because 7 exceeded the limit, 5 was as high as the results could go.

The *first number*, *step amount*, and *limit* can all be expressed as numbers (either integer or floating-point, positive or negative), variables that resolve to a number, or as calculations that evaluate to a number. They can also be characters (try it out for fun), but that's for another book.

The colon operator has four main uses:

- To define a vector. Example: v = 1 : 2.5 : 100
- To specify a range of array indexes. Examples: M(1:3) M(1:3, 2:4)
- To act as a placeholder for the full range of an array dimension. Example: M(:, 4)
- To reshape an array into a single column or two-dimensional array. Example: M(:)

This will make more sense if you see it in action, and the Command Window is just the place for that. First, we'll need some arrays to work with. Use the following commands to create sample arrays:

```
>> V = 1:5
>> M = randi(10, 4)
>> A = randi(15, 4, 5, 2)
```

You probably want more explanation of the randi options. The command, M = randi(10,4), means create a 4 x 4 matrix of random numbers between 1 and 10. The command, A = randi(15, 4, 5, 2) causes MATLAB to create a 3-dimensional matrix with 4 rows, 5 columns, and 2 pages, filled with random numbers between 1 and 15. These are just two examples of the many parameter sets that randi accepts. For more information, see the MATLAB help. The gist is that we now have a vector, a matrix, and a 3-D array to work with.

Enter this command, which asks for elements 3 through 5 in the vector V and stores them as the new variable named *sub*:

```
>> sub = V(3:5)
```

Next, we'll get a single element from the matrix M and then its upper-left 2 x 2 corner.

Enter this command:

```
>> M(2, 4)
```

Tip: *The step amount can be left out of the colon operation. For example firstnumber:limit as in 1:5.*

Tip: *The ability of a single function to accept a variable number and/or type of parameters is called overloading.*

Tip: *When used to create a vector, the colon operator always results in a row vector.*

You see the result:

```
ans =

    5
```

The command asked for the element in row 2, column 4 of matrix M. It returned a floating-point number. Enter this command:

```
>> M(1:2, 1:2)

ans =

     9    7

    10    1
```

That command asked for columns 1 and 2 from row 1 and row 2 of M. It returned a 2 x 2 matrix. The row is specified first and the column second as in *M(row, column)*.

Now to grab all of (and only) column 1. There are two ways to ask for it: Try this:

```
>> M(1:4, 1)

>> M(:, 1)
```

Tip: *The end keyword stands in for the last row or column. For instance A(1:end, 3) specifies column 3 from each row in matrix A.*

Putting a colon in place of any array dimension tells MATLAB to get the full range of that dimension – in this case all rows. It is equivalent to 1:end. You won't always know your array dimensions ahead of time (though they're easy enough to retrieve), so this can be an efficient way to generalize your code. M(:,:) would therefore return the entire array; though it's simpler to just say M.

The colon operator is extremely useful as a selector of elements, and you'll use it often in all kinds of situations. For example, create a new matrix made up of every other row of every other column in an existing array. Try this:

```
>> A1 = randi(15, 8, 10)

>> A2 = A1(1:2:8, 1:2:10)
```

Really look at A2 and understand where each number came from in A1 and why.

If you find it confusing, make two steps out of it. Try this:

```
>> A2 = A1(1:2:8, :)

>> A3 = A2(:, 1:2:10)
```

The colon-specified ranges can also be placed on the left side of an equation to allow you to assign a new value to only those selected array elements.

```
>> A1(1:2:8, 1:2:10) = 0
```

You can also replace those elements with elements from another array, as long as the second array has the dimensions of the replaced elements. For example, enter the following commands to replace the 2 x 2 center elements of a 4 x 4 matrix:

```
>> S = ones(10, 4)
>> T = randi(100,2)
>> S(2:3, 2:3) = T
```

Remember that the colon operator is simply a shortcut to express an array. The following two commands give the same result:

```
>> A2 = A1(1:2:8, 1:2:10)
>> A2 = A1([1 3 5 7], [1 3 5 7 9])
```

However, when you deal with the big data sets that you'll encounter in real-world problems, the compact form that the colon operator allows is often the only practical choice.

The following are a few examples of the use of the colon operator, matrices, and the *end* keyword to affect select portions of a matrix. Enter the following commands and observe their results:

```
>> X = ones(3,7)
>> X(2,[1 3]) = 2
>> X(2, 1:3) = 3
>> X([2 1], 2) = 4
>> X(end, 2) = 5
>> X(end, end) = 6
>> X([1 end], 4) = 7
>> X(1:end, 5) = 8
>> X(1, end-1) = 9
>> X(1:end, 5:6) = 10
>> X(1:end, 2:3) = [11 11; 12 12; 13 13]
>> X(:) = X(1,end)
```

You can use negative numbers as a step amount. The second of the following commands flips a matrix top to bottom.

```
>> X = magic(5)
>> X(end:-1:1, 1:end)
```

The final use of the colon operator is a little unexpected given how we've used it so far. This use reshapes an array rather than selects from it. Why would you want to do that? Well, for one example, there are functions such as sum() that operate only on the elements in a column vector. Enter this command in the Command Window:

```
>> M(:)
```

This command converts an array into a column vector. It takes each column in turn and adds it to the bottom of the previous column.

This can be applied to both vectors (turning a row vector into a column vector and a column vector into, well, the same column vector) and multi-dimensional arrays.

Tip: *The* ones() *function sets the array values to ones, in case that wasn't obvious.*

Tip: *You might read the line A2 = A1(1:2:8, 1:2:10) as "assign to A2 the values found in A1 rows 1 through 8, stepping by 2s (rows 1, 3, 5, 7) and from those rows use columns 1 through 10, stepping by 2s."*

Tip: *Remember, the values in the square brackets are the array indices. You might read the line A2 = A1([1 3 5 7], [1 3 5 7 9]) as "assign to A2 the values found in A1 row 1, columns 1, 3, 5, 7, & 9, and so on for rows 2, 5, and 7."*

Tip: *Remember, the comma (,) separates the row index from the column index.*

Tip: *Remember, the semicolon (;) separates rows in a list of matrix values.*

Tip: *The* magic(n) *function fills an n-by-n matrix with integers from 1 through n-squared so that every row and column adds up to the same sum. This is just a handy way of constructing some test values.*

Try these commands:

>> *V(:)*

>> *A(:)*

Want to look at your full array and maybe make some changes? This would be a good time to explore one of the capabilities of the Workspace. We'll use the array A1 that you declared above. The Workspace displays each of the variables that are active in the Command Window.

Workspace	
Name ▲	Value
A1	*8x10 double*
A2	*4x10 double*
A3	*4x5 double*
V	*[1,2,3,4,5]*

Figure 3.3 **Workspace Variables**

Data structures that are beyond the Workspace's capacity to display just report the size and type in the **Value** column. But double-click a variable's name, and that data is displayed in a spreadsheet-style **Variables** window. Double-click on variable A1 in the Name column to see its contents.

Note: Because A1 was created using the randi *function, your specific data may not be the same as seen here.*

	1	2	3	4	5	6	7	8	9	10
1	10	10	4	14	10	2	4	11	2	8
2	6	9	3	15	11	4	7	6	4	12
3	13	4	4	7	4	13	15	6	6	11
4	8	5	7	2	2	1	9	15	11	14
5	6	8	5	4	5	14	8	1	3	14
6	15	4	14	7	5	11	4	14	11	6
7	14	13	7	9	7	8	8	14	2	11
8	9	3	3	4	8	9	10	12	10	3

Figure 3.4 **Variables Window**

You can click in any cell and change the value. Right-clicking on the spreadsheet gives you the options to insert or delete whole rows and columns, among other options. Working in the Variables window can be more straightforward and efficient than manipulating this data through the Command Window's prompt.

Array Functions

Here are some functions that are useful when creating or working with an array. We'll discuss more array functions when we go deeper into matrices.

Function	Action
zeros	Create an array of all zeroes
ones	Create an array of all ones
rand	Create an array of random floating-point numbers

randi	Create an array of random integers
true	Create an array of logical ones
false	Create an array of logical zeroes
diag	Create a diagonal array, a square array with values on the diagonal from upper left to lower right and zeroes elsewhere
cat	Concatenate arrays
horzcat	Concatenate arrays horizontally, must have same number of rows
vertcat	Concatenate arrays vertically, must have same number of columns
length	Return the size of the largest dimension
size	Return the array's dimensions (as an array)
ndims	Return the number of the array's dimensions
numel	Return the number of elements in the array
isrow	Is the array a row vector?
iscolumn	Is the array a column vector?

Cell Arrays

Cell arrays are a special type of array that can contain more than one data type. Each individual *cell* (the cell array equivalent of an *element*) of the array can be any data type and of any size. Think of a cell array as a spreadsheet; in fact, if you import a spreadsheet into MATLAB, it becomes a cell array. The only difference as far as usage is concerned is that it is created using the cell() function and you use *curly brackets { }* to specify the index to the element rather than using parentheses. Because the original data is often supplied in a spreadsheet, cell arrays are used fairly often.

You declare the size of the cell array when you create it with the cell() function. Enter the following commands into the Command Window:

>> a = cell(4)

>> b = cell(4,5)

>> c = cell(4,5,2)

>> A = [4 5 2]

>> d = cell(A)

The first command, cell(4), creates a 4 x 4 cell array. The next, cell(4,5), creates 4 x 5 cell array. The next, cell(4,5,2), creates a three-dimensional cell array 4 x 5 x 2. The final command also creates a 4 x 5 x 2 cell array, but uses a vector to provide the dimensions. All of these cell arrays are initialized with empty arrays at each index as placeholders.

Next, try these examples showing different syntax used to access individual cells. We'll use the 4 x 5 cell array *b* that we defined earlier.

Note: *Unlike the other data types discussed up to this point, cell arrays aren't as common in other languages. You can sometimes doctor up something with the same functionality, but it's not often a built-in type.*

Note: *A cell array is actually an array of pointers, so technically it still follows the rule that says an array can only contain a single type. A pointer is just an address in the computer's memory. Each pointer in the array is the location in memory of that element. Pointers can make your head spin, and MATLAB doesn't allow you to see their value or manipulate them like you can in a language like C++, so just consider this an interesting implementation detail for now. If you didn't know it, you'd be just fine.*

>> *b{2,2} = 4.2*

Your Command Window shows results like that in Figure 3.5.

```
>> b = cell(4,5)
b =
  4×5 cell array
    []    []    []    []    []
    []    []    []    []    []
    []    []    []    []    []
    []    []    []    []    []
```

Figure 3.5   Result of cell array b(2,2) = 4.2

Continue with the following statements:

>> *b{1,1} = 'Good morning'*

>> *b{1,2} = "Bonjour"*

>> *b{3,4} = cos(45)*

>> *b{4,1} = int16(432)*

As you add each value to the cell array, MATLAB shows you the full array with its values of various types. Again, note the difference in the representation of the two string types.

Cell Array Functions

Here are some useful functions for cell arrays.

Function	Action
cell	Create a cell array
iscell	Is the variable a cell array?
cell2mat	Convert a cell array to a matrix. Only works if all elements are of the same type.
cell2struct	Convert a cell array to a structure
struct2cell	Convert a structure to a cell array

Structures

A struct (short for "structure") is another collection of various data types. *Structures*, as a native data type, are found in most programming languages. Each item in a struct is a named *field*, and each field has a value. Structures are great for gathering something's properties together, which is a convenient model of the real world. For example, you could create a *person* structure that contains fields for first name, last name, age, social security number, address, and criminal record. A field can be any data type: an integer, a floating-point number, a string, an array, or even another structure (that would work well for the criminal record). Our example structure will hold information about something more interesting – rockets.

There are several ways to create a structure, and MATLAB is very flexible about allowing you to create and modify the structure on the fly. We'll begin by defining an empty structure.

Enter the following assignment statement into the Command Window:

>> *rocket = struct*

This statement is a little unusual compared to anything we've seen before, because struct is a function call, but without a parameter list, even an empty one. The function simply returns an empty structure and the assignment statement gives it a name. You will learn more about functions in a later tutorial.

To add a field/value pair to the structure, use the dot notation like this: *structName.fieldName*.

>> *rocket.manufacturer = "SpaceX"*

If the field already exists, MATLAB assigns the new value to the field. If the field doesn't exist, MATLAB adds both the field and its value to the structure. In fact, if the structure itself doesn't exist, MATLAB creates both the structure and the field for you. Therefore, the rocket = struct line is optional here.

Now, add a few more rocket facts:

>> *rocket.model = "Falcon9"*

>> *rocket.height = 70*

>> *rocket.diameter = 3.7*

>> *rocket.stages = 2*

You now have a structure that holds a string and three floating-point numbers.

You can also declare a structure all at once by using the struct function with a comma-separated parameter list that alternates field names with their value: field1, value1, field2, value2, etc. Try this example:

>> *person = struct('name',"Ludwig",'age',20,'height',6.1)*

Note that field names must use single quotes, but values can use either single or double quotes.

A field also can be an array or another structure. Try this next.

>> *enginesStage1.quantity = 9*

>> *enginesStage1.burnTime = 162*

>> *enginesStage1.totalThrustAtSeaLevel = 7600*

>> *enginesStage1.totalThrustInVacuum = 8227*

>> *rocket.engines = enginesStage1*

You've just added a structure, called *engines*, to our existing structure *rocket*. To access a field of a structure within a structure, just continue the dot notation one more level.

>> *time = rocket.engines.burnTime*

>> *rocket.engines.totalThrustAtSeaLevel = 7607*

But wait! This rocket has engines in each stage. It would be great to have information about each one stored under *engines*. Well, as it happens, when a structure is created, it is actually a 1 x 1 structure *array*. To add another element to the array (another structure, that is) use a normal index subscript plus the dot notation. Give it a try now:

> *>> enginesStage2.quantity = 1*

> *>> enginesStage2.burnTime = 397*

> *>> enginesStage2.totalThrustAtSeaLevel = "N/A"*

> *>> enginesStage2.totalThrustInVacuum = 934*

> *>> rocket.engines(2) = enginesStage2*

Thanks to that last command, *engines* is now an array that contains two structures. Because *engines* is an array, each item it holds must be of the same data type. That both of the elements are **structs** isn't enough to meet that requirement – *each structure in the array must have the same fields*. To illustrate that, try this illegal assignment:

> *>> engine3 = struct*

> *>> engine3.burnTime = 250*

> *>> rocket.engines(3) = engine3*

The error message "Subscripted assignment between dissimilar structures." isn't the clearest way of saying "*this structure is not like the others,*" but that's what it means. The structure *engine3* has one of the same fields as the others, but is missing the rest. However, notice that the data type of the same field in different structures does not have to match: the *totalThrustAtSeaLevel* value is numerical for the first engine, but a String for the second engine.

Double-click the variable *rocket* in the Workspace and examine its contents in the **Variables** window, just as you did for arrays earlier. This will give you a visual representation of the internal layout of the structure.

To access a field in a structure within an array within a structure (yipes!), just add the index subscript to the dot notation:

> *>> time = rocket.engines(1).burnTime*

You can extrapolate this expansion of a single structure into an array to the *rocket* structure itself, adding a second set of rocket data. Try this:

> *>> rocket(2).model = "Gemini"*

Notice that MATLAB has created the rest of the structure too, with an empty array as a placeholder value for the other fields. After all, the structures in the array have to have the same fields, so there's no other way that this could be done legally. In the same way, when you add a new field to an existing structure in an array, that field is added to all of the other structures in the array as well, each with an empty array as a placeholder value. Enter this command:

> *>> rocket(1).engines(2).type = "Merlin"*

Examine the *rocket* variable in the **Variables** window. You'll see that rocket.engines(1) now has a *type* field too.

Structure Functions

Here are some useful functions when creating or working with structures.

Function	Action
struct	Create a structure as a 1 x 1 structure array
fieldnames	Return a cell array of all the fieldnames in the structure
rmfield	Remove a field from a structure, or from all structures in an array
struct2cell	Convert a structure to a cell array
cell2struct	Convert a cell array to a structure

Zoom! You have finished Tutorial 3.

Key Terms

column vector	*end*	*matrix*	*vector*
curly brackets { }	*field*	*row vector*	
element	*index*	*scalar*	

Key Commands

==	iscolumn	ones	zeros
cat	isempty	rand	
cell	ismatrix	randi	
cell2mat	isrow	rmfield	
cell2struct	isscalar	size	
diag	isvector	struct	
false	length	struct2cell	
fieldnames	magic	sum	
horzcat	ndims	true	
iscell	numel	vertcat	

Exercises

 Exercise 3.1

Using the colon operator, create a 1 x 10 array in which the first five numbers are the odd numbers between 1 and 9, inclusive, and the last five numbers are the even numbers between 2 and 10, inclusive. Hint: a comma is involved.

Exercise 3.2

Create a 6 x 6 array with the randi() function. Write an equation, using the colon operator, that retrieves the upper left 3 x 3 corner of the array.

Exercise 3.3

Create an 8 x 8 array with the randi() function. Write an equation, using the colon operator, that creates a new array from every other row and column in the original array.

Exercise 3.4

Create a structure called *student* that contains four fields: a character array called *name*, an int8 called *age*, a floating-point number called *GPA*, and a string array called *major*. Assign these fields any appropriate value. Convert the structure to a cell array and multiply each age entry times three.

Exercise 3.5

Using a combination of structure arrays and cell arrays, implement a Matlab structure called house with the following content:

```
house
  address
  year_built
  rooms
    bedroom
      200 sq_ft
      furniture: bed, dresser, nightstand
    kitchen
      100 sq_ft
      appliances: stove, refrigerator, dishwasher
    man/woman_cave
      300 sq_ft
      gaming_systems: Xbox, Playstation
    garage
      1000 sq_ft
    cars
      car1
        make
        model
        year
      car2
        make
        model
        year
```
Hint: Build from the smallest part to the largest.

PROGRAMMING BASICS: LOOPING AND CONDITIONALS

Introduction

In this tutorial we'll be talking about elements and concepts that are found in MATLAB, but also in most of the other programming languages. Up to this point we've discussed things that hold a value or values or, in the case of operators, perform an action on those values. *Loops* and *conditionals*, on the other hand, are concerned with the flow of code through your program. This means that to look them over, you need a program to flow through.

Before we dive in, a quick revisit to logical values, which are going to come up a lot in this tutorial. In the Command Window, MATLAB shows you the result of a logical expression as a 1 or 0 of type *logical*. These are equivalent to the constants **true** and **false**, respectively, and are available for you to use in your code instead of 1 or 0 if you choose. Because it is more intuitive to discuss conditional statements as evaluating to true or false, that's what we'll do here.

Using the MATLAB Editor

In this chapter, you will use loops and *branching* with conditional statements. So far you have written code in scripts, which are the most basic type of MATLAB program. A *function* is a package of code that can be called from the Command Window or from another function that allows you to request inputs and send outputs. In other words, a function performs a task and (usually) returns a value or values that you can assign to a variable. We'll be going into functions in detail in the next chapter, but we'll give you some basics here to get started.

MATLAB provides an editor for writing scripts and functions. Scripts and functions are stored in a *.m* file, which is a file type specific to MATLAB. Functions allow inputs to be passed in and outputs to be sent out. Scripts are essentially a list of commands that could be typed into the Command Window that have been grouped into a program file.

The editor can be launched in several ways. One way is to click New in the upper left of the ribbon and then choose Function from the drop-down menu as shown in Figure 4.1.

This opens a new, unnamed *.m* file in the Editor with the basic skeleton of a function provided, along with a bit of helpful text to remind you of its requirements. The ribbon switches to the Editor tab, with its set of commands and options.

Objectives

When you have completed this tutorial, you will be able to

1. Define the most basic of functions and save it in a .m file.

2. Declare and use the two loop styles found in MATLAB: while and for.

3. Declare and use the two conditional styles found in MATLAB: if(-elseif-else) and switch.

4. Understand the flow of execution in a function that contains looping and conditional statements.

5. Exit a function at the line of your choosing.

Figure 4.1

Here are a couple more ways you can begin creating a function.

- Type edit at the >> prompt in the Command Window. This opens a new, blank, unnamed file in the Editor and the ribbon (where you see Home, Plots, Apps, etc.) switches to show the Editor tab.

- Type edit *filename* at the prompt in the Command Window (substituting an actual file name for *filename*). This opens a new, blank file in the Editor using the name you provided and saves it as *filename*.m in your current directory. The ribbon switches to the Editor tab.

Tip: *You don't need to include the ".m" in the filename – MATLAB adds it automatically – but if you do, such as entering* edit loops.m, *MATLAB knows what you mean and names the file correctly.*

For now, we'll use that last method to keep things simple. In the Command Window, type this command at the prompt:

>> *edit loops*

*Select **Yes** at the prompt if asked whether to create the file*

The Editor window opens above the Command Window and you should see something like Figure 4.2. Note the file *loop.m* in the Current Folder window and that we've switched from the Home tab to the Editor tab.

The Editor tab
on the ribbon

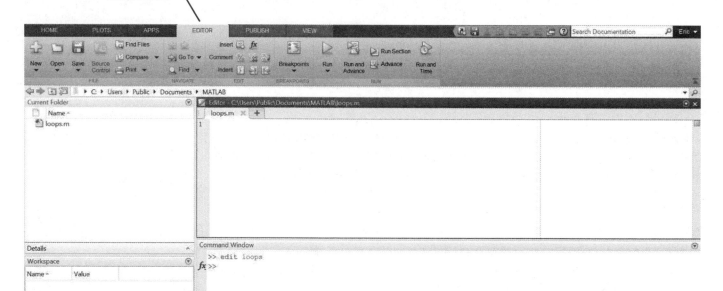

Figure 4.2   **The Editor Window with loop.m**

Enter the following lines into the Editor exactly as shown with a blank line before the word end:

function loops()

Tip: *You can also type the lines of code in any standard editor, like Notepad, and paste it into the MATLAB Editor later.*

```
Editor - C:\Users\fores\Documents\MATLAB\loops.m
 loops.m   +
1   function loops()
2
3 -   end
```

Figure 4.3   **The Function loops.m**

Loops

A *loop* enables you to execute the same block of code over and over, usually with at least one piece of different data each time. Here are some examples of how a loop is used:

- Perform the same operation on each member of a set

- Walk through a set of information looking for a match

- Perform an operation a given number of times

Assume you want to write a plan to add together the first five even numbers. Say that you don't know code yet, so you write a set of instructions in your native language. In programming, this is called an *algorithm*, and it's a good first step in attacking a programming problem. Here's what you might jot down:

1. Start with the first even number: 2

2. Get the second even number: 4

3. Add 2 + 4

4. Get the third even number: 6

5. Add (2 + 4) + 6

6. Get the fourth even number: 8

7. Add ((2 + 4) + 6) + 8

8. Get the fifth even number: 10

9. Add (((2 + 4) + 6) + 8) + 10

Notice that as of step 4, you essentially just repeat steps 2 & 3 over and over with different numbers. Now say that you want to write down the instructions for adding the first 1,000,000 even numbers. Write that out - we'll wait. No thanks? Rightly so. Let's generalize the procedure above.

1. Get a number

2. Get the next number

3. Add those numbers and call the sum X

4. Get the next number

5. X = X + the new number

6. Repeat steps 4 & 5 until you're done

7. Get on with your life

But how do you know when you're done? You need to set an end *condition*, which in this case is "Have I retrieved a million numbers yet?"

The condition you set will evaluate to true or false. But what if your condition was "Does X = 3?", which can never happen if you're adding even numbers. What if your condition can never be met? That is what's spoken of in hushed tones as an *infinite loop*. Sometimes, infinite loops are used on purpose, but more often they're a programming error. If you expect your function to return some information, but it just sits there churning, you can suspect that it's in an infinite loop. More on that later in the tutorial.

while Loops

The first of the two loop types in MATLAB is a while loop. It has this form:

```
while condition
        block of code
    end
```

The block of code is executed as long as the condition is true. Because the condition is evaluated at the beginning, it's possible for the block of code never to be executed at all.

The condition can be a simple logical statement, a very complex logical statement, or even a call to another function, as long as the result is true or false.

Next, we should talk about code formatting. A computer doesn't care what your code looks like; it's just a bunch of 0s and 1s to its cold, cold, silicon processor. The whole thing could be on one line and the result would be the same. But humans like things organized. As with informative variable names and use of white space, formatting your code using indentations to group statements can make your code easier to understand, both for you and for others. Indent code between an opening statement and an **end** statement (this applies to loops, branches, and entire functions). Loops and branches within those loops and branches are indented another level, and so on. MATLAB helps automatically place your cursor to line things up and indent appropriately.

Let's put a simple while loop in the function.

In the Editor window for the loops.m file, enter the lines of code for the function as shown in Figure 4.4:

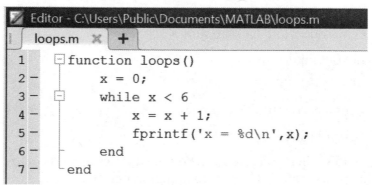

Figure 4.4 **while Loop**

The first line initializes the variable x with a value of 0. Next, the code enters the while loop. It first evaluates the condition, which is x < 6. The current value of x is 0. Because 0 is less than 6, the condition is true and the code that follows, up to the nearest **end** statement, is executed. We will call this the *loop body*. In our loop body, 1 is added to the current

Tip: Two or four spaces is the most common indentation. You can set your preference in the MATLAB preferences, in the Tab setting under Editor/ Debugger.

value of x and that new value is output to the screen using the **fprintf** command. Don't worry about the details of **fprintf** for now. (If you're curious, consult the MATLAB help.) At that point, the program flow returns to the while statement and it is evaluated again. Finally, when the value of x reaches 6, the while x < 6 statement evaluates as false, the while loop jumps to its end and runs the next line of code, which is the end statement for the function, therefore the function exits.

Next you will use the Command Window and enter the name of the function to call it. Calling a function executes its lines of code.

Click: in the Command Window area

>> loops

The result of that command should look like Figure 4.5.

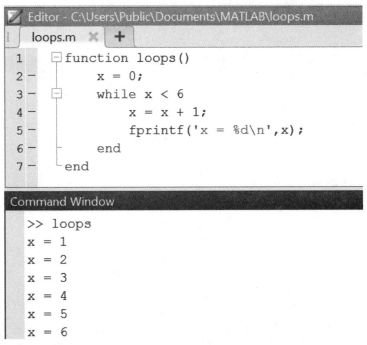

```
function loops()
    x = 0;
    while x < 6
        x = x + 1;
        fprintf('x = %d\n',x);
    end
end
```

Command Window
```
>> loops
x = 1
x = 2
x = 3
x = 4
x = 5
x = 6
```

Figure 4.5   Executed Function loops.m

Some programming languages, such as C++ and Java, require a semi-colon at the end of most lines to say "this is the end of the line," and will result in an error if one is absent. In MATLAB, they are optional but encouraged in a function or script. As in the Command Window, a semi-colon at the end of the line causes that line to execute silently rather than sending its result to the Command Window (which can make output long and confusing).

As an experiment, remove the semi-colon after x = x + 1 *and run the function again. And that's just one line...imagine watching hundreds of lines of code do that as your function executes. Put that semi-colon back in place.*

MATLAB is also somewhat unusual in that the loop condition does not need to be enclosed in parentheses, as is the case in many languages. You're free to use parentheses if you'd like, however.

To sum the first million even numbers, add the following code to your **loops** function in the Editor window before that last **end** statement:

Tip: *You'll notice that until you put the semi-colon on an assignment statement, the equal sign is highlighted in yellow. If you hover your cursor over that equal sign, MATLAB displays a pop-up message suggesting that you add the semi-colon. You can click the Fix button in that message and MATLAB will do the job for you.*

Also keep an eye out for items underlined with a small red squiggle. If you see one, hover for a message from MATLAB about an issue that it has detected. For instance, remove the closing parenthesis in the fprintf line and then click elsewhere on the page to cause the squiggle to appear with a pop-up message about your now-mismatched parentheses.

Tip: *Some languages, but not MATLAB, have a variant of the while loop that's called a do-while loop. The difference lies in when the condition is evaluated. In a while loop, it is evaluated at the start; in a do-while loop, it is evaluated at the end. The effect is that the body of a while loop might not be executed at all, but the body of a do-while loop will always be executed at least once.*

Tip: *In the fprintf command the formatting option %d prints the value as a signed integer and \n starts a new line.*

Tip: *Don't be concerned when you see red lines under code as you type. These error notifications will often resolve when you add the final parenthesis in or use your variable in a function.*

```
count = 0;
runningSum = 0;
evenNum = 0;
while count < 1000000
        evenNum = evenNum + 2;
        runningSum = runningSum + evenNum;
        count = count +1;
end
fprintf ('The sum is %d/n', runningSum);
```

so that you end up with this:

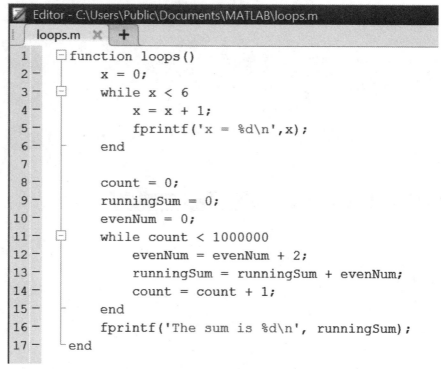

Figure 4.6

> ***Click in the Command Window to activate that area and enter the following to run your function again.***

```
>> loops
```

Now you have your answer for the sum of the first million even numbers. You should see the result 1000001000000. In the blink of an eye, your loop has run one million times. This is the awe-inspiring power of loops.

for Loops

The other type of loop in MATLAB (and countless other languages) is a for loop. A for loop uses a loop index to control how many times the block of code is executed. The loop index is often also used as a variable within the block of code.

Here is the basic form of a for loop.

```
for loopIndex = firstValue : lastValue

    block of code

end
```

The *loopIndex* can have any legal variable name. There's an old programming tradition of using 'i' as a loop variable, but this is just a convention. Some people think it's too hard to distinguish an i from the number 1 or a small L. Use any variable name within reason.

The *firstValue* is the first index value; *lastValue* is the upper limit of the index values. They can be integers or floating-point numbers. These can be given as numbers, as variables, as the result of an operation or a call to another function, or as any mix of those things. Did you notice the use of the colon operator to separate them? Type that part of the **for** command in the Command Window:

>> *index = 1:4*

The variable *index* is assigned the array [1 2 3 4]. This is precisely what's happening in the **for** command – absolutely normal colon operator behavior, including the default increment of 1 between each array member. (Review the last tutorial if you don't recall the details.)

No matter how you supply *firstValue* and *lastValue*, the result is an array of indexes. Try it for yourself by entering these commands into the Command Window and observing the values of *index*:

>> *num = 12;*

>> *firstValue = 2;*

>> *lastValue = 8;*

>> *index = firstValue : lastValue*

>> *index = ceil(-2.4) : 5*

>> *index = 3/8 + num : 2*7.5*

Next, you will open a new function file and try out a **for** loop.

In the Command Window, type the following command and hit [Enter]:

>> *edit forloops*

*Select **Yes** at the prompt if asked whether to create the file*

The *forloops.m* file is created in your current directory and is opened in the Editor as a blank file.

Next, you will enter some lines of code into the Editor. This code outputs the iteration of the loop, the loop index, and the value of the variable *x* as it proceeds through the loops.

On your own, enter the lines of code into the Editor as shown in Figure 4.7.

Note: *The for condition can be any array, including a 3D or larger array. As always, see the Help page on* for *loops for more information as you need it.*

```matlab
1   function forloops()
2       x = 10;
3       loopCounter = 1;
4       for i = 1:3
5           fprintf("Entering loop #" + loopCounter);
6
7           loopCounter = loopCounter + 1;
8
9           fprintf("\n\ti (loop index) = " + i);
10          fprintf("\n\tx (on entry) = " + x);
11
12          x = x + i;
13
14          fprintf("\n\tx (on exit) = " + x);
15          fprintf("\n");
16      end
17  end
```

Figure 4.7   for Loop

Tip: *Yes, the name of the function **does** have to match the name of the file.*

In the Command Window, enter the function name to execute the function:

>> *forloops*

The output that you see in the Command Window should look like this:

    Entering loop #1
        i (loop index) = 1
        x (on entry) = 10
        x (on exit) = 11
    Entering loop #2
        i (loop index) = 2
        x (on entry) = 11
        x (on exit) = 13
    Entering loop #3
        i (loop index) = 3
        x (on entry) = 13
        x (on exit) = 16

Note the correlation between the loop number and the loop index. That's a result of the default index increment of 1.

What if you want the loop index to increase by some amount other than 1 each time? You can use the standard array notation that we've seen before to insert a step value between *firstValue* and *lastValue*, each piece separated by colons as shown in the next block of example code:

```
for loopIndex = firstValue : stepAmount : lastValue
    block of code
end
```

Try it out by changing your code in the Editor.

*Click: **Editor and change line 4 from***

> *for i = 1:3*
>
> *to*
>
> *for i = 1 : 0.3 : 3*

In the Command Window, enter the name to execute the function:

> **>> forloops**

Notice that due to the new step value, the *lastValue* 3 won't be part of the loop index array. The command tells MATLAB to create an array that starts at 1, increase each subsequent value by 0.3, and stop once the value is greater than the *lastValue*. Verify this by entering this command in the Command Window and observing the resulting array:

> **>> i = 1 : 0.3 : 3**

The *stepAmount* can be negative if you want to count down instead of up. Just make sure that your *firstValue* is greater than your *lastValue* or the loop body will never execute. Test out this next bit of code in the Command Window:

> **>> x = 10**
>
> **>> for i = 2 : -0.4 : 1**
>
> **>>   x = x + 1;**
>
> **>> end**

Like the other parts of the array declaration, the *stepAmount* can be a number, a variable, or the numerical result of a calculation. It can be an integer, or it can be a floating-point number.

So, if the loop condition always evaluates to an array, why not just use an array? Why not, indeed; it's perfectly valid.

> **Go back to your forloops code in the Editor and change the line**
>
> *for i = 1 : 0.3 : 3*
>
> *to*
>
> *for i = [1, 1.3, 1.6, 1.9, 2.2, 2.5, 2.8]*

Again go to the Command Window and enter the function name to execute it:

> **>> forloops**

The result is the same as the last run, just using a different form of the same loop indexes. In this case, one form is less tedious to type, but if you want to use an array with more than a few members, the colon operator is more practical.

Tip: *The spaces around the colons are optional. They are there for clarity, but you can leave them out.*

Tip: *You can leave the ; off the line x = x + 1 and watch the results as variable x increments. This helps you get the idea of the loop and is useful in debugging.*

Tip: *The commas are optional, they just make it easier to read.*

You can also assign the array to a variable and use the variable in the for statement as follows:

```
indexArray = 1:200;
for i = indexArray
```

Conditional Statements

This class of statements is used by MATLAB to decide whether to execute a single block of code or to decide between multiple blocks of mutually exclusive code. These statements come in two flavors: the *if* statement (with a couple sub-flavors of its own) and the *switch* statement.

if Statements

The basic form of an *if* statement is simple:

```
if condition
    block of code
end
```

Tip: *The condition doesn't have to be enclosed in parentheses as is common in other languages, but you can use them if you'd like.*

The condition is any expression that evaluates to a logical value, just like a **while** loop. If the condition is met (that is, the expression evaluates to true) then the block of code is executed. If the condition is not met (it evaluates as false) then the block of code is skipped and program jumps to the next line after the **end** statement.

Returning to the terminal velocity problem from the first tutorial, if it were a function rather than a script, you could make the code more flexible by adding an **if** block to set your gravity and atmospheric density variables based on a String variable called *planet*.

```
if (strcmp(planet, 'Mars'))
    gravity = 3.711;
    atmosphericDensity = 0.020;
end
```

if-else Statements

Sometimes, you might find yourself with two blocks of code where one and *only* one of them must execute. In that case, you can use *if-else* construction like this:

```
if condition
    block of code 1
else
    block of code 2
end
```

In the **if-else** form, if the condition is met, block 1 is executed. If the condition is not met, block 2 is executed. One or the other will always run. You can only have one **else** block per **if** statement.

Use the Command Window to create a new function to test some if-else statements.

>> *edit branches*

You should have *branches.m* open in Editor. Enter the lines of code as shown in Figure 4.8:

```
loops.m    forloops.m    branches.m    +

1   function branches()
2       planet = 'Earth';
3       if strcmp(planet, 'Mars')
4           gravity = 3.711;
5           atmosphericDensity = 0.020;
6       else
7           gravity = 9.798;
8           atmosphericDensity = 1.2170;
9       end
10      disp("Gravity = " + gravity);
11  end
```

Figure 4.8   Code in the Editor for Branches.m Function

On your own, execute the branches function in the Command Window and observe the gravity setting.

Because the condition is false, the **else** block of code is used and the values set to those of Earth.

Now go back in and change the value of planet *from* 'Earth' *to* 'Mars' *and run the function again.*

You'll see that it now outputs Mars' gravity.

if-elseif-else and if-elseif Statements

The *if-else* form can be expanded further by including one or more *elseif* blocks, each with their own condition:

```
if condition 1
    block of code 1
elseif condition 2
    block of code 2
elseif condition 3
    block of code 3
...
else
    block of code 4
end
```

Note: The disp *command is another output command like* fprintf, *but simpler.*

Tip: *An ellipsis (…) in a general definition like the* if-elseif *statement, means that you can keep adding statements beyond the couple that are shown as illustration. The ellipsis is just a placeholder. You might also see it in a code example. There, it means that some code has been omitted for clarity.*

As in the **if-else** version, only one block of code, the first that meets its condition, is executed.

The following flow chart shows how code execution proceeds through an **if-elseif-else** statement, depending on whether each condition evaluates to true or false:

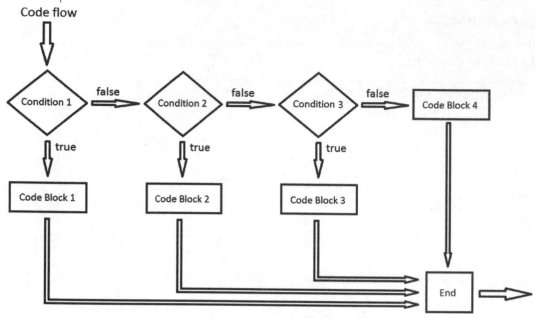

Figure 4.9   **Flow Chart for an if-elseif-else Statement**

You can have all the **elseif** blocks you need, though if you have too many, you might examine your design to look for a more elegant solution.

The **else** block is an *optional* fallback in case no other condition is met. If you omit the **else** block, you allow for the possibility of no block of code at all being executed. The coding problem you're solving at the time will tell you which option best suits your needs.

switch Statements

When you have more than a few alternate blocks of code which require a long series of **if-elseif** statements, a more elegant solution can be the *switch*. A **switch** statement is used when the same condition is being tested throughout, but for different values (called *cases*). Its general form is as follows:

```
switch variable
    case value 1
        block of code 1
    case value 2
        block of code 2
    case value 3
        block of code 3
    ...
    otherwise
        block of code 4
end
```

Tip: *Programmers describe some code as elegant. As with any aesthetic property, there is some vagueness about what defines it, but in general it's brevity, clarity, and efficiency. Keep in mind there is no single right answer to a code solution; there are many ways to get from point A to point B. Doing it cleverly (but not so cleverly that others can't understand and modify your code) and in a way that uses minimal resources and executes quickly should be one of your design goals whenever you face a new programming challenge.*

To illustrate how this works, let's look again at the terminal velocity example. Using a form of conditional statement, the variables for each planet's gravity and atmospheric density can be set based on a String variable that specifies the planet, defaulting to Earth in the absence of a match. This solution can be coded in either of two ways: a **switch** statement or an **if-elseif-else** statement. Both have precisely the same result.

Switch	if-elseif-else
switch planet	if strcmp(planet, 'Mars')
case 'Mars'	gravity = 3.71;
gravity = 3.71;	atmosphericDensity = 0.020;
atmosphericDensity = 0.020;	elseif strcmp(planet, 'Saturn')
case 'Saturn'	gravity = 10.44;
gravity = 10.44;	atmosphericDensity = 0.19;
atmosphericDensity = 0.19;	elseif strcmp(planet, 'Neptune')
case 'Neptune'	gravity = 11.15;
gravity = 11.15;	atmosphericDensity = 0.45;
atmosphericDensity = 0.45;	else
otherwise	gravity = 9.798;
gravity = 9.798;	atmosphericDensity = 1.217;
atmosphericDensity = 1.217;	end
end	

Let's give each of these a try using a script similar to that we made for calculating the terminal velocity of a 25 mm ball bearing.

On your own, clear the Command Window and Workspace

 Use the Open file button to open terminalVelocityConditionals.m from the data files

Tip: *Don't try to import it. It is a script, not a data file.*

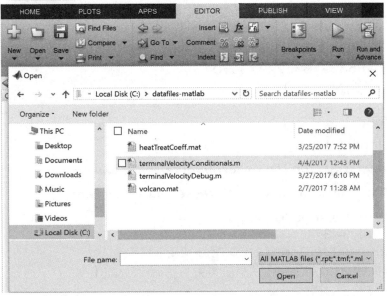

Figure 4.10

Tip: *Notice that this is a script and not a function. It does not contain a* function *statement. You will learn more about the differences between scripts and functions in the next tutorial.*

Tip: *Notice that if this script contained a* clear *statement, that would clear the variable list, including the planet value set from the Command Window. Then, the variable planet would not be defined and the* switch *statement wouldn't work.*
You'd have the same problem if you forgot to set the planet variable at all.

Scroll through and examine the code in the Editor

Case statements have been added to the terminal velocity script for you. See Figure 4.11.

A comment has been added reminding you to set the variable *planet* equal to one of the cases. In a later tutorial you will learn how to create a function that prompts for the information. For now, you will use the Command Window to set the *planet* variable.

```
Editor · C:\datafiles matlab\terminalVelocityConditionals.m
terminalVelocityConditionals.m  ✕  +
1    % program to calculate terminal velocity
2    % Vt = terminal velocity
3    % dia = diameter of sphere (in cm)
4    % mass = mass of the falling object(65.4710 grams)
5    % gravity = acceleration due to gravity (in cm/sec^2)
6    % atmosphericDensity = density of the medium through which the object is
7    % falling (in g/cm^3)
8    % area = cross-sectional area of the object,
9    %          for a sphere it is pi*radius^2
10   % C = drag coefficient (sphere is approx 0.47)
11   planet = 'mars'
12   switch planet
13     case 'Mars'
14       gravity = 371;
15       atmosphericDensity = 0.000020;
16     case 'Saturn'
17       gravity = 1044;
18       atmosphericDensity = 0.00019;
19     case 'Neptune'
20       gravity = 1115;
```
```
Command Window
fx >>
```

Figure 4.11   Terminal Velocity Script Using Case Statement

Click: in the Command Window and make the following entries:

>> *planet = 'Neptune'*

>> *terminalVelocityConditionals*

The script executes. Results display in the Command Window as shown:

Vt =

   1.185869183622224e+04

cm/s units

planet =

   'Neptune'

On your own, save the file as terminalVelocityCond2.m

Replace the case statements with these if-elseif statements:

```
if strcmp (planet, 'Mars')
   gravity = 371;
   atmosphericDensity = 0.000020;
elseif strcmp (planet, 'Saturn')
   gravity = 1044;
   atmosphericDensity = 0.00019;
```

Tip: *Notice the values are in different units than in the earlier examples.*

```
elseif strcmp (planet, 'Neptune')
    gravity = 1115;
    atmosphericDensity = 0.00045;
else
    gravity = 981;
    atmosphericDensity = 0.001217;
end
```

On your own, execute the new script in the Command Window by typing terminalVelocityCond2

As you see, the results are the same. These are parallel methods, including the **otherwise** block being optional. There is no obvious winner at this scale. The **switch** block is a bit cleaner to read, but using **if** and **elseif** works just fine. Pick the best option based on the situation (or whichever you like better, which is a perfectly valid reason at times).

Taking this example further, each planet could have its own **case, if,** or **elseif**. A good use of the **otherwise/else** block would be as a data check on your *planet* variable, outputting a statement to the effect that Pluto, the Moon, Krypton, or Tatooine aren't real planets, so try again.

Now let's use loops and conditionals together. You might be surprised at what a sophisticated function you can put together already.

Loops and Conditionals: Make a Negative Example

Clear the Workspace and the Command Window.

Use the Command Window to enter the following line:

>> *edit negative*

This function turns an image file into its negative. Each pixel in an image is simply a small, colored dot. That color is determined by a mixture of three color values: red, green, and blue. Therefore, the full description of a pixel requires its x and y position in the image and its three color values. This function uses the MATLAB function **imread**, which creates a 3-dimensional array that represents the image pixel-by-pixel: (x, y, red), (x, y, blue), and (x, y, green).

More importantly, we're showing you how loops and conditionals are used and interact in a real-world application.

Here is the function code for you to enter into the *negative.m* file. We've included a comment for most lines, explaining what they do. You can leave those out of the file if you want. We have also provided this as a data file in case you don't want to type it all, though you will learn more and for a longer time by typing it.

```
function negative()
    % Specify the image file
    imageFile = 'c:\data\pretty.png';
    % Read the pixel information into an array
    ImageMap = imread(imageFile);
    % Get the height and width of the image
```

```
11 -   planet = 'mars'
12 -   if strcmp(planet, 'Mars')
13 -       gravity = 371;
14 -       atmosphericDensity = 0.000020;
15 -   elseif strcmp(planet, 'Saturn')
16 -       gravity = 1044;
17 -       atmosphericDensity = 0.00019;
18 -   elseif strcmp(planet, 'Neptune')
19 -       gravity = 1115;
20 -       atmosphericDensity = 0.00045;
21 -   else
22 -       gravity = 981;
23 -       atmosphericDensity = 0.001217;
24 -   end
25 -   dia = 2.5;
26 -   area = pi*((dia/2)^2);
27 -   mass = 65.4710;
28 -   C = 0.47;
```

Tip: *You may need to switch to run the script from the folder locations where you have it stored.*

Tip: *Well-commented code is a sign of a fine programmer. It can be an enormous help to anyone who examines the code later, including yourself. You don't have to comment every line, but commenting at important points is a good practice.*

Tip: *If you do not have the data files for the book stored in c:\datafiles-matlab, or you don't have write permissions there, then copy pretty.png to a folder where you do have write permissions and update the line of code:*
imageFile = 'c:\data\pretty.png';
to reflect the path where you have stored the file.

```
rows = size(ImageMap,1);
columns = size(ImageMap,2);
% Ensure that the image is valid
if (rows > 0 && columns > 0)
    y = 1;
    % Process one row at a time
    while y <= rows
        % Look at each column in that row
        for x = 1:columns
            % Get each red, green, and blue value
            % and change the value to its opposite
            for i = 1:3
                ImageMap(y, x, i) = 255 - ImageMap(y, x, i);
            end
        end
        % Go to the next row and do it all over again
        y = y + 1;
    end
    % Write the information back to the image file
    imwrite(ImageMap, imageFile);
else
    disp("The image is invalid");
end
end
```

Figure 4.14　　　pretty.png

Figure 4.15　　　Negative

On your own, save the negative.m function.

Use your file browser or other tools to view the image pretty.png.

Run the negative function in the Command Window (you can also use the Run button on the Editor ribbon).

Now, view the image pretty.png again.

Be sure to look at the image before you run the function. After you run the negative function, it should be a negative (opposite colors).

Run the negative.m function again

Now it will be back to normal (the negative of a negative is a positive).

This function might normally be written with nested for loops rather than a for loop inside a while loop. We did it that way just to show that any of these elements can contain any of the other elements. Feel free to nest a switch inside a for inside a while inside an if; whatever is the best solution to the coding problem you're trying to solve.

Exiting a Loop

If you're looping through a large data set, there's no point in continuing to loop once you've found what you're looking for or otherwise accomplished your task. There are three commands that let you exit a while loop or a for loop. The difference between them is where the program goes once the loop is exited.

break

The break statement stops execution of the loop and goes to the statement following the loop's end statement.

continue

The continue statement stops the execution of that specific loop iteration and goes to the next iteration.

return

The return statement exits the function altogether. This statement isn't just for loops; you can put a return command anyplace in your function where you've determined that the function has done its job. We'll go into more detail about the return command in a later tutorial.

Next, you will create a new function called loops2 to explore the break and continue commands.

>> edit loops2

On your own, add the following lines to that file in the Editor:

```
function loops2()
   for i=1:10
      disp(i);
   end
end
```

Run the loops2 function using the Command Window on your own.

The numbers 1 through 10 are displayed.

On your own, edit the code as follows to add an if statement inside the for loop:

```
function loops2()
   for i=1:10
      if i == 6
         break;
      end
      disp(i);
   end
end
```

On your own, run the loops2 function in the Command Window.

Now only numbers 1 through 5 are displayed. Once the variable i became 6, the if condition became true, its break command was executed, and the loop exited without going on to 10.

Next, replace the break command with the continue command and run the function again.

On your own, edit the code as follows:

```
function loops2()
    for i=1:10
        if i == 6
            continue;
        end
        disp(i);
    end
end
```

On your own, run the new loops2 function in the Command Window.

If you look carefully at the list of numbers, you'll notice that 6 is now missing. Once the variable i became 6, the if condition became true, the continue command was executed so that the rest of that loop iteration (the output statement disp(i)) was skipped, i became 7, and the loop continued to completion.

Stopping an Infinite Loop

If you find your function stuck in an infinite loop, you'll need to halt that function before you can fix the problem. A portion of the Editor tab's ribbon is a section titled Run. This section presents different actions depending on whether the function is currently executing. It allows you to run the function directly instead of typing the command in the Command Window. When the function is running, it displays several actions; the one that's important to us now is *Pause*.

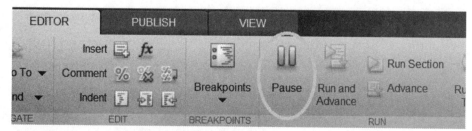

Figure 4.12   Quit Debugging Button

Hit Pause when you want to temporarily halt the function. The Run section then becomes the Debug section. We'll talk about this section in detail when we talk about troubleshooting. For now, all we need is the *Quit Debugging* button.

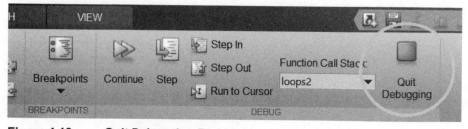

Figure 4.13   Quit Debugging Button

Click Quit Debugging to close and exit the function and return you to the Editor.

Create a new function file called loops3.

Put these lines into loops3:

```
function loops3()
    while 1
        disp("All work and no play makes Jack a dull boy");
    end
end
```

This loop will run forever because the condition can never be anything but true (logical(1)).

Click: **Run** button from the Editor ribbon to launch **loops3.**

Click: **Pause**

Click: **Quit Debugging** to get out of the loop and leave the function

You'll use this method of bailing out for reasons other than infinite loops, too, so be sure to keep it in your mental toolbox. Think about the loops you are coding and plan a graceful exit. If you notice the code running on and on, be ready with the Pause button.

You have finished Tutorial 4.

Key Terms

branching	firstValue	lastValue	loops
condition	function	loop body	pause
conditionals	infinite loop	loop index	

Key Commands

case	for	imwrite	while
else	if	otherwise	
end	imread	switch	

Exercises

Exercise 4.1

Explain why this loop will never execute, in terms of an array declaration and the resulting array.

```
for i = 1 : -2 : 10
    disp(i);
end
```

Exercise 4.2

Rewrite this **while** loop as a **for** loop.

```
x = 1;
while x <= 10
    disp(x);
    x = x + 1;
end
```

Exercise 4.3

Rewrite this **if** statement as a **switch** statement. Assume that the String variable *hero* has been defined earlier in the function.

```
if strcmp(hero, 'Superman')
    villain = 'Bizarro';
elseif strcmp(hero, 'The Thing')
    villain = 'Fin Fang Foom';
elseif strcmp(hero, 'Hellboy')
    villain = 'Baba Yaga';
else
    villain = 'Dracula';
end
```

Exercise 4.4

The factorial of a number is the product of all of the positive integers less than or equal to the number. An exclamation point (!) after a number is used to designate the factorial of that number, for example 150!. Create a script to calculate the factorial of 150 using loops and conditionals. MATLAB has a built-in function to compute factorials, but do not use it in your code. Look in Help for the factorial function, and use it to check your results.

Exercise 4.5

Create your own code example that will result in an infinite loop. Modify the code to provide a conditional that will exit the loop.

MATRICES

Introduction

You can use MATLAB to write a program that is as complex and useful as any in another programming language, as well as easily visualize the data and create 3D graphs, but MATLAB's ability to deal with matrices sets it apart.

The branch of mathematics called *linear algebra* is built on the interaction and manipulation of matrices to solve complex problems. Linear algebra is generally taught following the calculus series. Linear algebra and matrix mathematics are used in engineering (modeling circuits, robotic motion), physics (quantum mechanics), computer graphics (reflections, 3D projections to 2D), and digital photography (device-based color correction), just to scratch the surface. We can't teach linear algebra here, but you'll get a glimpse of some of its potential uses and gain an understanding of matrices. When you have a need for solutions using matrices, you'll find MATLAB invaluable.

What Is a Matrix?

As you learned previously, a *matrix* (plural: *matrices*) is a two-dimensional, rectangular array consisting of rows and columns. A *vector* is a matrix with only one row (1 x N) *or* one column (N x 1). A *scalar* is a single value, but is seen by MATLAB as a matrix with just one row *and* one column (1 x 1). Matrices are often shown using square brackets:

$$\begin{bmatrix} 1 & 2 & 3 \\ 4 & 5 & 6 \end{bmatrix}$$

When used in this book, this bracketed form is for illustration purposes only. You can't enter those all-enclosing brackets in MATLAB commands, nor does MATLAB use them to show you a matrix. However, brackets *are* used to indicate a matrix to MATLAB. These brackets can be entered directly into the Command Window or used in code.

```
[1 2 3; 4 5 6]
```

See that semi-colon between 3 and 4? A semi-colon in a MATLAB matrix definition is a *row separator*. 1, 2, 3 is the first row; 4, 5, 6 is the second. Each row must contain the same number of elements.

Each element is referred to by its position in the array as (row, column). Notice that we're back to parentheses to specify the index. **Square brackets are for defining matrices, parentheses for using matrices.**

In MATLAB it's **always row first, then column**. That's how you enter the position and that's how MATLAB shows the position to you.

Objectives

When you have completed this tutorial, you will be able to

1. Explain the difference between an array, a matrix, a vector, and a scalar.

2. Specify an array in MATLAB using brackets and semi-colons.

3. Understand the concepts of a diagonal, identity, and magic matrix.

4. Add, subtract, multiply, and divide a matrix by a scalar or vector.

5. Raise the values in a matrix to a scalar power or a vector of powers.

6. Use MATLAB functions to create test matrices.

7. Seed a sequence of random numbers so that the sequence can be replicated.

8. Recognize when two matrices meet the requirements for matrix multiplication.

9. Perform a matrix multiplication manually.

10. Have a rudimentary understanding of an inverse matrix and a matrix determinant.

Tip: *Individual values in a row can be separated by either commas or spaces. [1,2,3; 4,5,6] is the same as [1 2 3; 4 5 6]. It's down to personal preference. In this book, we'll generally omit the cluttering commas.*

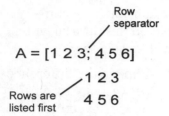

Row separator

A = [1 2 3; 4 5 6]

Rows are listed first

1 2 3
4 5 6

A is a 2 x 3 matrix: 2 rows, 3 columns

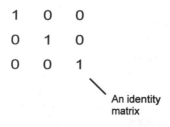

An identity matrix

Enter the following commands in the Command Window:

> *>> A = [1 2 3; 4 5 6]*

> *>> size(A)*

The **size** function returns **ans =   2   3**, telling you that A has 2 rows and 3 columns. Rows and columns, always in that order.

There are several specialty matrices. Here are three of the simpler ones:

Diagonal matrix

A *diagonal matrix* is a *square matrix* with the special feature that the diagonal elements from the upper left to the lower right corner contain values while all other elements in the matrix are 0.

The **diag** function creates an N x N diagonal matrix, using a set of N values on the diagonal. Try **diag** in the Command Window:

> *>> V = [1:4];*

> *>> A = diag(V)*

Matlab returns the following:

```
A =

     1    0    0    0
     0    2    0    0
     0    0    3    0
     0    0    0    4
```

The **diag** function can also retrieve the diagonal from a matrix, returning it as a vector. Try it next:

> *>> X = diag(A)*

Matlab returns the following:

```
X =

     1
     2
     3
     4
```

Identity matrix

An *identity matrix* is a special type of diagonal matrix which contains all zeros, except for a diagonal of ones running downward from the upper left corner. The identity matrix is the matrix version of the number 1; multiplying a matrix A by its identity matrix results in matrix A.

You can create an identity matrix using the **eye** function. Use the **eye** function to create a 4 x 4 identity matrix:

> *>> eye (4)*

Matlab returns the following identity matrix:

```
ans =
    1    0    0    0
    0    1    0    0
    0    0    1    0
    0    0    0    1
```

Magic matrix

A *magic matrix* is a square matrix of size N x N, made up of the integers 1 through N^2. Each row and each column adds up to the same sum. Try to do *that* by hand! Try the **magic** function next:

>> *magic (5)*

```
ans =
   17   24    1    8   15
   23    5    7   14   16
    4    6   13   20   22
   10   12   19   21    3
   11   18   25    2    9
```

Tip: *When you supply only a single number N to a matrix-generation function, you're asking for a square matrix of size N x N.*

Matrix Math

Matrix math is any math where at least one of the elements in the equation is a matrix. You can add or subtract a scalar from each matrix element. You can multiply or divide each matrix element by a scalar. You can raise each element in a matrix to a power. Matrices with the same dimensions can be added together or subtracted from one other.

Matrices also can be multiplied and divided, but there are two very different types of multiplication and division!

Creating Test Matrices

You're going to need some matrices to test out some matrix math functions in the next pages. You can enter them yourself, but there are easier ways. Here are a few functions, some of which we've already mentioned, that make it easy to create test matrices:

magic(*n*): Creates a square array of size *n* x *n* made up of the numbers 1 through n^2. Each row and column adds up to the same number.

eye(*n*): Creates a square array of size *n* x *n* made up of all zeros, with the exception of a diagonal of ones from the upper-left corner to the lower-right.

rand(*n*) or rand(*n,m*): Creates an array of size *n* x *n* or *n* x *m* made up of random floating-point numbers between 0 and 1.

randn(*n*) or randn(*n,m*): Creates an array of size *n* x *n* or *n* x *m* made up of small random floating-point numbers with a mean of 0, which is to say that it includes both negative and positive numbers.

randi(*max,n*) or randi(*max,n,m*): Creates an array of size *n* x *n* or *n* x *m* made up of random integers between 1 and *max*.

ones(*n*) or ones(*n,m*): Creates an array of size *n* x *n* or *n* x *m* made up entirely of ones.

zeros(*n*) or zeros(*n,m*): Creates an array of size *n* x *n* or *n* x *m* made up entirely of zeros

A Few Words on Random Number Generation

There really is no such thing as a truly *random number,* and doubly so where computers are concerned. Instead, *pseudo-random* numbers are generated through various functions designed to look like they're producing random output. Each new number is based on the number before it, with the first number being based on a *seed* provided by the function. If, instead, you provide the seed number, the sequence of "random" numbers will be the same every time. Let's prove it by entering a few commands in the Command Window:

> *>> randi (10,1,10)*
>
> *>> randi (10,1,10)*

First, we hope that you pressed ↑ to repeat the command rather than keying it in twice. Second, notice that the second output is entirely different than the first. Now enter these commands:

> *>> rng (0)*
>
> *>> randi (10,1,10)*
>
> *>> rng (0)*
>
> *>> randi (10,1,10)*

Now your "random" numbers should be the same each time. We got:

$$9 \quad 10 \quad 2 \quad 10 \quad 7 \quad 1 \quad 3 \quad 6 \quad 10 \quad 10$$

The command rng (0) resets the *random number generator*, so you get the same seed every time, and therefore the same sequence of numbers. The ability to reproduce a "random" sequence of numbers turns out to be a good thing because it allows you to use consistent data when you're testing your code.

Array Math Operations

Array math operations are the most straightforward of matrix math operations, because they are performed on each element separately. To perform array math operations, *the arrays involved must have the same dimensions.* These operations apply to matrices, but they can also apply to three-dimensional and higher arrays. Review the Arrays and Structures tutorial as a refresher for array basics.

Matrix Addition and Subtraction

You can add two or more matrices together or subtract one matrix from another. Again, the rule is that all matrices in the equation must have the same dimensions (same number of rows and columns). Each element in one matrix is matched with the corresponding element at the same position in the other matrix (or matrices) and the operation is applied. For *matrix addition* with matrix A and B, A(1,1) is added to B(1,1), A(1,2) is added to B(1,2), etc. Here it is visually:

$$\begin{bmatrix} a & b & c \\ d & e & f \end{bmatrix} + \begin{bmatrix} h & i & j \\ k & l & m \end{bmatrix} = \begin{bmatrix} a+h & b+i & c+j \\ d+k & e+l & f+m \end{bmatrix}$$

Matrix subtraction works in the same way. Notice that order counts, just as it does when you subtract one scalar from another: A – B is not the same as B - A. (If you remember your grade school math terms, this says that subtraction is not *commutative*.)

$$\begin{bmatrix} a & b & c \\ d & e & f \end{bmatrix} - \begin{bmatrix} h & i & j \\ k & l & m \end{bmatrix} = \begin{bmatrix} a-h & b-i & c-j \\ d-k & e-l & f-m \end{bmatrix}$$

Try it out in the next steps. Because you will be using random number generators, your values will not exactly match the examples shown.

```
>> A = randi (10,3,5)
   A =
       2    5    5   10    9
      10    9   10    7   10
      10    2    8    1    7
>> B = randn (3,5)
   B =
       1.0347   0.2939  -1.1471  -2.9443  -0.7549
       0.7269  -0.7873  -1.0689   1.4384   1.3703
      -0.3034   0.8884  -0.8095   0.3252  -1.7115
>> C = rand (3,5)
>> D = [1 2 3; 4 5 6; 7 8 9]
   D =
       1    2    3
       4    5    6
       7    8    9
>> E = A + B + C
>> F = A – C
>> G = A - D
   "Matrix dimensions must agree."
```

Tip: *Keep in mind you will have your own set of random numbers. Because of that not all of the output is shown for these examples.*

Tip: *As that last command shows, MATLAB will tell you if your matrix dimensions differ.*

Multiplication and Division

As long as at least one dimension agrees, *array multiplication and division* work on an *element-by-element* basis. However, we need to introduce three important new operators. They are *element-wise multiplication* (.*), *element-wise division* (./), and *element-wise power* (.^) .

The Element-wise (Dot) Operators

The period before these operators makes them element-wise operators rather than matrix operators. We'll get into matrix multiplication and division in a bit. For now, just know that matrix multiplication and division don't operate on an element-by-element basis like these dot operators.

The multiplication (.*), division (/*), and power (.^) operators have the same precedence as their non-dotted counterparts.

$$\begin{bmatrix} a & b & c \\ d & e & f \end{bmatrix} .* \begin{bmatrix} h & i & j \\ k & l & m \end{bmatrix} = \begin{bmatrix} a*h & b*i & c*j \\ d*k & e*l & f*m \end{bmatrix}$$

$$\begin{bmatrix} a & b & c \\ d & e & f \end{bmatrix} ./ \begin{bmatrix} h & i & j \\ k & l & m \end{bmatrix} = \begin{bmatrix} a/h & b/i & c/j \\ d/k & e/l & f/m \end{bmatrix}$$

Here's an example that shows the element-wise multiplication of two matrices that agree in only one dimension:

$$\begin{bmatrix} a & b & c \\ d & e & f \end{bmatrix} .* \begin{bmatrix} g & h & i \end{bmatrix} = \begin{bmatrix} a*g & b*h & c*i \\ d*g & e*h & f*i \end{bmatrix}$$

Scalar Math Operations

When one of the operands in an array equation is a scalar, that scalar value is applied to each element. For instance, adding a scalar *n* to a matrix (or a higher-dimensional array) adds *n* to each element in the matrix.

$$n + \begin{bmatrix} a & b & c \\ d & e & f \end{bmatrix} = \begin{bmatrix} n+a & n+b & n+c \\ n+d & n+e & n+f \end{bmatrix}$$

Enter these commands in the Command Window:

```
>> A = ones(5)
>> A + 6
>> B = rand(1,3)
>> x = 1
>> B + x
```

Your results should be similar to the following (but not exactly the same as random numbers were used to generate B.)

```
A =
   1   1   1   1   1
   1   1   1   1   1
   1   1   1   1   1
   1   1   1   1   1
   1   1   1   1   1
ans =
   7   7   7   7   7
   7   7   7   7   7
   7   7   7   7   7
   7   7   7   7   7
   7   7   7   7   7
B =
   0.4387   0.3816   0.7655
x =   1
ans =
   1.4387   1.3816   1.7655
```

Subtraction works in the same way. Again, order counts.

$$n - \begin{bmatrix} a & b & c \\ d & e & f \end{bmatrix} = \begin{bmatrix} n-a & n-b & n-c \\ n-d & n-e & n-f \end{bmatrix}$$

$$\begin{bmatrix} a & b & c \\ d & e & f \end{bmatrix} - n = \begin{bmatrix} a-n & b-n & c-n \\ d-n & e-n & f-n \end{bmatrix}$$

Multiplication and division of a matrix by a scalar give the expected element-by-element results whether you use the dot operators or not. However, you should think about the process and use the one that reflections your intent.

$$\begin{bmatrix} a & b & c \\ d & e & f \end{bmatrix} .* n = \begin{bmatrix} a*n & b*n & c*n \\ d*n & e*n & f*n \end{bmatrix}$$

$$\begin{bmatrix} a & b & c \\ d & e & f \end{bmatrix} ./ n = \begin{bmatrix} a/n & b/n & c/n \\ d/n & e/n & f/n \end{bmatrix}$$

Try these commands:

```
>> A = magic(3)
   A =
       8   1   6
       3   5   7
       4   9   2
```

Tip: *While both the array and matrix form of the multiplication and division operators give the same result in this case, it is bad form to make that assumption. Use the correct operator for what you mean to do.*

```
>> n = 5
>> A / n

    ans =
      1.6000    0.2000    1.2000
      0.6000    1.0000    1.4000
      0.8000    1.8000    0.4000
>> A ./ n

    ans =
      1.6000    0.2000    1.2000
      0.6000    1.0000    1.4000
      0.8000    1.8000    0.4000
>> n ./ A

    ans =
      0.6250    5.0000    0.8333
      1.6667    1.0000    0.7143
      1.2500    0.5556    2.5000
>> n / A

      Error using  /
      Matrix dimensions must agree.
```

Aha! You can divide a scalar by a matrix using array division (which results in an array), but not using matrix division. Why that is will become clear soon. The takeaway is that when you want to multiply or divide an array element-by-element, put a period in front of the operator. This also applies if you want to raise each element to a power; in that case, use the .^ operator.

Element-wise Division by a Scalar

The general case of the element-wise division of a matrix by a scalar looks like this:

$$\begin{bmatrix} a & b & c \\ d & e & f \end{bmatrix} ./\, n = \begin{bmatrix} a/n & b/n & c/n \\ d/n & e/n & f/n \end{bmatrix}$$

Under the hood, when one of the operands is a scalar, MATLAB expands that scalar to be an array of the same dimensions as the array in the equation, with that scalar value at each position. Therefore, there's nothing special about applying a scalar to an array – it's just two arrays. That's why you can divide a scalar by an array and wind up with an array.

$$\begin{bmatrix} a & b & c \\ d & e & f \end{bmatrix} ./\, n = \begin{bmatrix} a & b & c \\ d & e & f \end{bmatrix} ./ \begin{bmatrix} n & n & n \\ n & n & n \end{bmatrix} = \begin{bmatrix} a/n & b/n & c/n \\ d/n & e/n & f/n \end{bmatrix}$$

Transposition

Transposition is a common matrix operation that flips the matrix so that rows become columns and columns become rows. The transposition operator is a period plus a single quote (.') and has a precedence secondary only to parentheses. A second transposition operator omits the period ('). If your matrix does not contain complex numbers (numbers that include the *imaginary number i*), the effects of the two transposition operators are the same.

Another way to accomplish the same operation is to use **transpose(A)**. This is handy when you find the ' and .' just too small to notice.

$$A = \begin{bmatrix} a & b & c \\ d & e & f \end{bmatrix} \quad A' = \begin{bmatrix} a & d \\ b & e \\ c & f \end{bmatrix}$$

Transposition of a row vector results in a column vector and vice versa. To demonstrate transposition, enter the following commands into the Command Window:

>> *A = [1:2:6; 3:5:15; 10:10:30]*

Result	A =		
	1	3	5
	3	8	13
	10	20	30

>> *A'*

Result	ans =		
	1	3	10
	3	8	20
	5	13	30

>> *A.'*

Result	ans =		
	1	3	10
	3	8	20
	5	13	30

Due to the lack of imaginary numbers, the two transpositions are the same. Now let's try some results with an imaginary number. One of MATLAB's strengths is the ability to work with imaginary numbers. They have a special symbol, *i* in MATLAB results. You can also type *i* after a value to indicate the imaginary portion. We will use the square root of -2 to produce our imaginary value.

Tip: *Remember when using the colon operator to specify a row as is done here, you must have the same number of elements in each row and in each column. If not, you will see the message, "Dimensions of matrices being concatenated are not consistent." The matrix doesn't have to be square (say 3 x 3), but it cannot contain rows with different numbers of elements.*

Tip: *Remember comma is the optional separator between elements in a row, and semi-colon is the separator between rows. You can leave out the commas and just use a space, but you need the semi-colons to indicate the end of that row.*

Try the next commands.

>> *format short*

>> *A = [1 3 10; 3 sqrt(-2) 20;   5 13 30]*

The result for matrix A should be (note the imaginary number portion):

```
A = 1.0000 + 0.0000i   3.0000 + 0.0000i  10.0000 + 0.0000i
    3.0000 + 0.0000i   0.0000 + 1.4142i  20.0000 + 0.0000i
    5.0000 + 0.0000i  13.0000 + 0.0000i  30.0000 + 0.0000i
```

>> *A'*

```
    ans =

        1.0000 + 0.0000i    3.0000 + 0.0000i    5.0000 + 0.0000i
        3.0000 + 0.0000i    0.0000 - 1.4142i   13.0000 + 0.0000i
       10.0000 + 0.0000i   20.0000 + 0.0000i   30.0000 + 0.0000i
```

>> *A.'*

```
    ans =

        1.0000 + 0.0000i    3.0000 + 0.0000i    5.0000 + 0.0000i
        3.0000 + 0.0000i    0.0000 + 1.4142i   13.0000 + 0.0000i
       10.0000 + 0.0000i   20.0000 + 0.0000i   30.0000 + 0.0000i
```

The transposition results are different and result in different signs for the imaginary number. You aren't likely to use imaginary numbers at the level this book is intended for, but if you do, keep in mind that .' and ' are not the same operator.

Here's an example of transposition used in an array multiplication. Enter the following in the Command Window:

>> *A = [1:6; 3:8; 4:9]*

```
    A =

        1   2   3   4   5   6
        3   4   5   6   7   8
        4   5   6   7   8   9
```

>> *B = [2 4 6]*

```
    B =

        2   4   6
```

>> *B .* A*

```
    Matrix dimensions must agree.
```

>> *B .* A'*

```
    ans =
             2   12   24
             4   16   30
             6   20   36
             8   24   42
            10   28   48
            12   32   54
```

The first multiplication results in an error because it attempts to multiply a 1 x 3 row vector by a 3 x 6 matrix. The second multiplication flips the matrix so that now it's multiplying a 1 x 3 row vector times a 6 x 3 matrix. Each row element of the matrix is multiplied by its corresponding element in the vector.

Raising to a Power

Matrix elements can be raised to a power through the power operator (.^). The form follows the other element-wise array operations:

$$\begin{bmatrix} a & b & c \\ d & e & f \end{bmatrix} .^\wedge 2 = \begin{bmatrix} a^2 & b^2 & c^2 \\ d^2 & e^2 & f^2 \end{bmatrix}$$

$$\begin{bmatrix} a & b & c \\ d & e & f \end{bmatrix} .^\wedge \begin{bmatrix} r & s & t \end{bmatrix} = \begin{bmatrix} a^r & b^s & c^t \\ d^r & e^s & f^t \end{bmatrix}$$

$$\begin{bmatrix} a & b & c \\ d & e & f \end{bmatrix} .^\wedge \begin{bmatrix} r \\ s \end{bmatrix} = \begin{bmatrix} a^r & b^r & c^r \\ d^s & e^s & f^s \end{bmatrix}$$

$$\begin{bmatrix} a & b & c \\ d & e & f \end{bmatrix} .^\wedge \begin{bmatrix} g & h & i \\ j & k & l \end{bmatrix} = \begin{bmatrix} a^g & b^h & c^i \\ d^j & e^k & f^l \end{bmatrix}$$

Raising a scalar value to an n x m matrix results in an n x m matrix.

$$2 .^\wedge \begin{bmatrix} a & b & c \\ d & e & f \end{bmatrix} = \begin{bmatrix} 2^a & 2^b & 2^c \\ 2^d & 2^e & 2^f \end{bmatrix}$$

Logical Operations

An *element-by-element logical operation* ($<, <=, >, >=, ==, \sim=, \&\&, \|$) can be performed on a matrix. For an n x m matrix, the result is an n x m matrix of logical values. Try this example:

>> A = [1 2 3 4 5 6; 3 4 5 6 7 8; 4 5 6 7 8 9]
    (or use your current matrix A)

>> B = A <= 5

The result

```
    B =

      3×6 logical array
        1  1  1  1  1  0
        1  1  1  0  0  0
        1  1  0  0  0  0
```

As with most of the other array operations, you can compare

- two matrices of the same size
- a matrix and a vector that share one of the matrix dimensions
- a matrix and a scalar

Matrix Math Operations

There are no separate matrix versions of addition and subtraction. There is a matrix version of the power operator, but it performs a linear algebra operation that involves concepts called eigenvalues and eigenvectors and won't be discussed in this book. Matrix multiplication and division are quite different operations than their array versions.

Matrix Multiplication

Multiplying a matrix by another matrix requires that **the number of columns in the first matrix is equal to the number of rows in the second matrix**. The result is a combination of multiplication and addition. To a newcomer this can seem like number manipulation for its own sake, without real-world application, but in fact it's the real power of matrices. This is what allows you to solve parallel equations simultaneously and run test-modeling operations on massive data sets.

Let's start with the conditions that make a matrix multiplication possible. Assume two matrices A and B. A has dimensions (3,5) and B has dimensions (5, 2). Here is the matrix multiplication equation, using the multiplication symbol you know:

```
A * B
```

Look at the size dimensions for the matrix:

```
(3,5)   (5,2)
```

Notice that the *inner* dimensions are the same value (5), so this is a valid matrix multiplication. The *outer* dimensions have no limitations.

This operation is not commutative. Order matters!

```
B * A -> (5,2)  (3,5)
```

The inner dimensions (2 and 3) do not match, so if you attempted this matrix multiplication you would see an error telling you so.

Back to A * B, the inner dimensions of the two matrices match. The result size is based on their outer dimensions, here (3,2).

Summing that up, for a matrix A of dimension (r,s) and matrix B of dimension (t,u):

- Matrix multiplication can only be performed if s = t
- The resulting matrix is of size (r,u)

Before we dive into the specifics of the operation, know that there's nothing difficult or new involved. It's just multiplication and addition. It can be confusing at first, though, so keep reading.

The general formula for matrix multiplication A * B is this:

$$R(r,c) = \sum_x A(r,x)B(x,c)$$

In words, the formula says that in the result matrix R, an element at location (r,c) is the summation (indicated by the Greek letter Σ or sigma) over each value of x of the products of the specified elements of matrix A and matrix B. This will make a lot more sense if you see an example. We'll use these two matrices:

$$A = \begin{bmatrix} a & b \\ c & d \\ e & f \end{bmatrix}$$

$$B = \begin{bmatrix} g & h \\ i & j \end{bmatrix}$$

Because the number of columns in A equals the number of rows in B, this is a legitimate matrix multiplication. The result matrix R will have dimensions (3,2). It's easiest to visualize the procedure by arranging the matrices like this, with result matrix R at the intersection:

```
    B
A   R
```

To multiply A x B, think of beginning with this:

Matrix B

Matrix A

Result matrix R

The value of R(1,1) = (a*g) + (b*i). Put another way, R(1,1) = A(1,1) * B(1,1) + A(1,2) * B(2,1). Think of it as this intersection:

R(1,1) = (a*g) + (b*i)

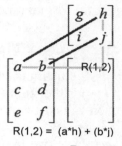

The value of R(1,2) = (a*h) + (b*j), or
R(1,2) = A(1,1) * B(1,2) + A(1,2) * B(2,2).

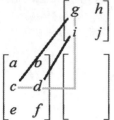

The value of R(2,1) = (c*g) + (d*i), or
R(2,1) = A(2,1) * B(1,1) + A(2,2) * B(2,1).

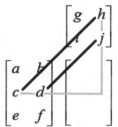

Starting to see the pattern?

The value of R(2,2) = (c*h) + (d*j), or
R(2,2) = A(2,1) * B(1,2) + A(2,2) * B(2,2).

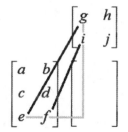

The value of R(3,1) = (e*g) + (f*i), or R(3,1)
= A(3,1) * B(1,1) + A(3,2) * B(2,1).

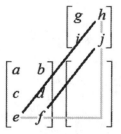

And finally, the value of R(3,2) = (e*h) +
(f*j), or R(3,2) = A(3,1) * B(1,2) + A(3,2) *
B(2,2).

Putting this all together, the result matrix R is as follows. Look for the patterns in the coordinates.

$$\begin{bmatrix} A(1,1)*B(1,1)+A(1,2)*B(2,1) & A(1,1)*B(1,2)+A(1,2)*B(2,2) \\ A(2,1)*B(1,1)+A(2,2)*B(2,1) & A(2,1)*B(1,2)+A(2,2)*B(2,2) \\ A(3,1)*B(1,1)+A(3,2)*B(2,1) & A(3,1)*B(1,2)+A(3,2)*B(2,2) \end{bmatrix}$$

Of course, with MATLAB this is all done for you in a couple of keystrokes, but now you understand what it's doing, which is essential. This is much different than multiplying each value times a scalar, or multiplying each element in a matrix times the same element in a different matrix. For those, use the dot operators. Linear algebra was invented long before computers, much less MATLAB, so having a computational tool that allows you to perform this operation accurately on large matrices is a splendid thing.

Matrix Division

Even though MATLAB provides an operator (/) to do so, technically there is no such defined operation as matrix division. There are three ways that the / operator is used by MATLAB in an equation of the form x = B/A, and they depend on the nature of A. Here, we quote directly from the MATLAB help:

- If A is a scalar, then B/A is equivalent to B./A.

- If A is a square n-by-n matrix and B is a matrix with n columns, then x = B/A is a solution to the equation x*A = B, if it exists.

- If A is a rectangular m-by-n matrix with m ~= n, and B is a matrix with n columns, then x = B/A returns a least-squares solution of the system of equations x*A = B.

We discussed the first point under array operations, so nothing new there. The last point is a linear algebra procedure beyond the scope of this book. The middle point is what we want to look at, as it is needed to solve many matrix problems. To understand its use, we need to learn about *inverse matrices*.

Inverse Matrices

Assume matrix A, matrix B, and matrix C, which are connected through this equation:

```
A * B = C
```

We know the values in matrices A and C but need to solve for matrix B. Using standard math, divide both sides by A, or, put another way, multiply each side by the inverse of A, just as 3 ÷ 4 is the same as 3 * ¼ (except that, being a matrix multiplication, order matters). The result is this:

```
B = C/A
```

Or, more specifically:

```
B = A⁻¹ * C
```

A^{-1} is called an *inverse matrix*. An inverse matrix must be an $n \times n$ square matrix. Its definition is $A * A^{-1} = A^{-1} * A = I$, where I is the $n \times n$ identity matrix. It's not precise, but as a shorthand you can think of $A*A^{-1}$ as equal to 1 in many calculations; it essentially cancels out.

An inverse matrix's value involves the matrix's *determinant*, and its *adjugate*. We're not going into those concepts in depth, but so that you have a basic idea of what they are, here's how to calculate them for a 2 x 2 matrix A.

$$A = \begin{bmatrix} a & b \\ c & d \end{bmatrix}$$

The *determinant* is a scalar value calculated as (a*d) – (c*b) for a 2 x 2 square matrix

The *adjugate* is a matrix, formed by swapping *a* and *d* and multiplying *b* and *c* by -1:

$$\begin{bmatrix} d & -b \\ -c & a \end{bmatrix}$$

With **det(A)** representing the determinant of matrix A and **adj(A)** the adjugate of matrix A, the formula for the inverse matrix of A is shown here:

$$A^{-1} = \frac{1}{\det(A)} \; .* \; adj(A)$$

Calculating the inverse of anything larger than a 2 x 2 matrix involves many more steps and falls under the "beyond the scope of this book" heading, plus MATLAB can do all of the above for any matrix with its **inv** function, which calculates an inverse matrix. If you just can't wait to get a deeper understanding, there are many good tutorials online that will both walk you through these calculations and demonstrate why it's better to use a computer than do them by hand.

Examples

Hey, remember when we used to talk about MATLAB? Time to get back to that and show you a brief example of using this matrix math.

MATLAB provides built-in functions for most of the operations we've discussed so far.

Function	Result
det (A)	Determinant of a matrix
inv (A)	Inverse of a matrix

For our first example, consider these ordinary linear equations.

```
3x + 3y = 16
2x - 5y = -4
```

Is there a value for x and y that satisfies both equations? To solve that problem, first we will represent the equations on the left as a matrix A that contains the coefficients and the values on the right as a matrix B. We want to solve for the variables on the left as a matrix D.

Give it a try in the next steps at the MATLAB Command Prompt.

Tip: *The variable and solution matrices, consisting of only one column, are mathematically known as column vectors.*

```
>> A = [3 3 ; 2 -5]

    A =

        3    3

        2   -5
>> B = [16; -4]

    B =

      16

      -4
```

We need the inverse of the coefficient matrix A, which MATLAB gives us through the inv function.

```
>> C = inv (A)

    C =

      0.2381    0.1429

      0.0952   -0.1429
```

To solve for D, multiply the matrix C by matrix B, which results in this calculation:

```
>> D = C * B

    D =

      3.2381

      2.0952
```

Therefore, the values x = 3.2381 and y = 2.0952 satisfy both equations, and the point (3.2381, 2.0952) is the intersection in a graph of the two lines represented by the original equations:

```
3x + 3y = 16

2x - 5y = -4
```

This method can also show when there is no solution (the lines never intersect) when MATLAB returns the *Inf* (infinity) constant.

As an exercise, lets check the values in the original equations. D(1) is the first element in solution matrix D, and D(2) the second.

```
>> 3*D(1) + 3*D(2)

    ans =

      16
>> 2*D(1) - 5*D(2)

    ans =

      -4.0000
```

The equations calculate correctly. Now let's try another problem.

A mighty battle against the forces of evil has left the Justice League's facilities in piles of rubble and space junk. They need to rebuild, so they're conducting a fundraiser for a new space station, a new Fortress

Tip: *In a later tutorial on symbolic math, you will learn about the* fplot *function. When you do, come back to this set of equations and plot the lines on a graph to see that their intersection is as we've calculated. For now, you could, for each equation, plug in a couple x values, solve for y, and graph. Two points for each equation make a line, so you should at least see that the intersection point is about where we said it would be.*

of Solitude and a new Batcave. In the first week of the fundraiser, donations were collected as follows:

- For the space station, each hero contributed $800, each sidekick $300 and each villain in custody $200 (call it restitution), for a total of $8900.

- For the Fortress of Solitude, each hero gave $700, each sidekick $400 and each villain $200, for a total of $9400. Superman is a good guy and people like him.

- For the Batcave, each hero gave $300, each sidekick $500 and each villain $100 for a total of $7700. Batman doesn't really go out of his way to make friends.

Assuming that each hero, sidekick, and villain gave money for each structure, how many heroes contributed? How many sidekicks? How many villains?

We could start by putting the information into algebraic equations, but instead we'll make a table that contains the data. This gives you a different way of thinking about a matrix, the information it holds, and how to visualize a matrix problem.

	Hero	Sidekick	Villain	Total
Space Station	800	300	200	8900
Fortress of Solitude	700	400	200	9400
Batcave	300	500	100	7700

Our coefficient matrix consists of the Hero, Sidekick, and Villain columns from that table. Enter the following command into the MATLAB Command Window.

>> *A = [800 300 200; 700 400 200; 300 500 100]*

The answer matrix is the Total column. Enter this command into the Command Window:

>> *C = [8900; 9400; 7700]*

Recall the general equation we're solving:

```
A * B = C
```

B, the unknown matrix, is our variables:

$$\begin{bmatrix} heroes \\ sidekicks \\ villains \end{bmatrix}$$

This time we'll take the inverse and do the multiplication in one step. Enter the following into the Command Window:

```
B = inv(A)*C
```

And there's your answer:

$$\begin{bmatrix} heroes \\ sidekicks \\ villains \end{bmatrix} = \begin{bmatrix} 6.0000 \\ 11.0000 \\ 4.0000 \end{bmatrix}$$

6 heroes, 11 sidekicks, and 4 villains contributed to the fundraising. Those sidekicks are really trying to look good and a lot of villains are still on the loose.

Many additional matrix functions are available; this chapter is just the tip of the iceberg. As usual, read more about the matrix functions and see many great examples in the MATLAB help files.

You have finished Tutorial 5.

Key Terms

array division	element-wise multiplication	matrix addition	scalar
array multiplication	element-wise power	matrix math	seed
commutative	identity matrix	matrix subtraction	square matrix
diagonal matrix	imaginary number i	pseudo-random	transposition
element-by-element	magic matrix	random number	vector
element-by-element logical operation	matrices	random number generator	
element-wise division	matrix	row separator	

Key Commands

.'	diag	rand	size
.*	eye	randi	transpose
.^	magic	randn	zeros
/*	ones	rng	

Exercises

 Exercise 5.1

A.*A results in the square of each value in matrix A. A*A, however, which gives an entirely different answer, can only be performed if what condition is true?

Exercise 5.2

In the second week of the Justice League fundraiser discussed in the lesson, contributions were as follows:

- For the space station: from each hero $500, each sidekick $300, and each villain $200: a total of $7100

- For the Fortress of Solitude: from each hero $1000, each sidekick $100, each villain $300: a total of $8300

- For the Batcave: from each hero $200, each sidekick $200, each villain $0 (really, nobody likes Batman): a total of $3400

Prove that the number of heroes, sidekicks, and villains is the same as found in the lesson (6 heroes, 11 sidekicks, 4 villains).

Exercise 5.3

Create a 3 x 3 matrix A, with any numerical content. Use MATLAB to determine its inverse, then *manually* perform the matrix multiplication A * A^{-1} to prove that the answer is 1.

Exercise 5.4

In Germany, annual professional fees are set by law for engineers, doctors, and lawyers. In Frankfurt, the total fees charged in a single year by 23 doctors, 18 lawyers, and 12 engineers were equivalent to $76,500,000 US. In that same year in Berlin, 25 doctors, 20 lawyers, and 16 engineers charged $85,500,000. In Munich, 18 doctors, 9 lawyers, and 20 engineers charged $55,000,000.

In MATLAB, use the formula B = A^{-1}*C to determine the annual fee charged by each doctor, each lawyer, and each engineer. Tip: Use 76.5 rather than 76,500,000 (etc.) in constructing that matrix.

Exercise 5.5

Using the matrices A and B from Exercise 5.4, in a single MATLAB command, calculate a new 3 x 3 matrix D that contains the specific amounts charged by each profession in each city. For example, (1,1) will be the total fees charged by doctors in Frankfurt, (2,2) the total fees charged by doctors in Berlin, and so on.

Exercise 5.6

To the matrix D created in Exercise 5.6, add a 4th column that is a sum of each row. Do this with a single MATLAB command. Hint: the sum function, the colon operator, and MATLAB help are your friends.

Exercise 5.7

To the matrix D from Exercise 5.7, add a 4th row that is a sum of each column. Do this with a single MATLAB command.

Exercise 5.8

Using a single MATLAB command applied to matrix D from Exercise 5.8, calculate a matrix E where each element is reduced by half to account for the local and national taxes paid by each profession.

Exercise 5.9

Based on matrix E from Exercise 5.9, and once again using a single MATLAB command, calculate a new 1 x 3 matrix F that specifies the average annual after-tax income of a doctor, a lawyer, and an engineer in this year. Are you still happy with your career choice?

Exercise 5.10

On Monday, a farmer sells 8 pounds of potatoes, 3 pounds of beets, and 6 pounds of asparagus. He receives a total of $44. On Tuesday, he receives $48 dollars for 5 pounds of beets, 3 pounds of potatoes and 7 pounds of asparagus. On Wednesday, he receives a total of $31 for 9 pounds of potatoes, 1 pound of beets, and 4 pounds of asparagus.

Using MATLAB, determine the inverse of the coefficient matrix. Manually perform the matrix multiplication to determine what price per pound the farmer charges for each of the produce types. Verify your answer using MATLAB.

FUNCTIONS AND SCRIPTS

Introduction

It's time for some serious coding! Here's where we put it all together: variables, loops, conditionals, matrices, and everything else we've talked about in the earlier tutorials. We'll look at two methods of complex data manipulations: *scripts* and *functions*.

We'll also review some of the concepts that we've talked about in those earlier tutorials, so don't be surprised if you feel a little déjà vu.

Comment Statements

First, let's revisit comment statements. A *comment* is a non-executable statement that can be included in either a script file or a function file. To enter a comment, begin the line with a percent sign (%).

Since you can use a comment in a script file, it follows that you can enter comments in the Command Window. That's true, but there's no real reason to do it. Nevertheless, try it here and you will see that comments are not executed. Enter the following at the >> prompt:

```
>> % This is a comment
>> % The following equation will not be executed:
>> % x = 100 / sin(52)
```

Comments are for humans, not for MATLAB. MATLAB skips right over them. Comments can explain a variable's intent, label a section, or otherwise explain your thinking; anything to make the code clear to yourself and others. You'll be surprised at how fast you'll need those reminders. Comments can go anywhere in the file and you can have as many as you need.

Bottom line: include comments! You'll thank yourself later and so will anyone else who has to read your code.

Scripts

You have created scripts already in these tutorials, but we'll review here and add more information. At its most basic, a *script* file is simply a convenience that allows you to avoid typing the same set of commands into the Command Window over and over. Instead, you can automate a set of commands by saving them in a script file and then executing them all by calling the script by name. In MATLAB, both script files and function files have a *.m* extension. Essentially, a script is a program.

Scripts can contain loops and conditionals. As of MATLAB R2016b, they can also contain their own helper functions, which can be called only from within the script. A script can easily be converted into a function, showing that in many cases, they're very similar.

Objectives

When you have completed this tutorial, you will be able to

1. Understand the importance of commenting your code.

2. Create a script from Command Window statements.

3. Prompt for values from the user.

4. Understand the difference between a script and a function.

5. Know the advantages of building a complex program from single-task functions.

6. Write an algorithm as a first step to writing code.

7. Understand how to return and accept multiple values from a function.

8. Know the difference between a main function and a local function.

9. Understand variable scope.

10. Use global and persistent variables.

11. Write a recursive function.

Tip: *You can use loops and conditionals in the Command Window too, but it's a little awkward. You must hit [Enter] between each line, but you won't be returned to the command prompt until you type the last end statement, at which point the block of code executes immediately. To run it again, you can select all the statements in the code block from the Command History, but you might as well make it into a script.*

Tip: *The Symbolic Math toolbox has the symunit feature for tracking and converting units. You will learn about the Symbolic Math toolbox in a later tutorial.*

Note: *Trigonometric functions in MATLAB are found in two forms: one that accepts radians (sin, cos, tan, etc.) and another that accepts degrees (sind, cosd, tand, etc.), so you could use the appropriate version instead of converting, but this provides an example. The takeaway is to know your units before you choose your trigonometric function!*

The essential thing to know about a script is that it operates in the *scope* of the Command Window, using the base Workspace. It can contain any valid command that you can execute in the Command Window, including calling functions or other scripts. It can recognize and act on variables that you've already defined (listed in the Workspace) or variables that it defines (which are added to the Workspace).

There are two types of scripts in MATLAB: regular scripts and Live Scripts (which have a *.mlx* extension). Live Scripts are discussed in a later tutorial. You can see many of them in action in the MATLAB help.

Script Example

Perhaps you want to calculate the magnitude of torque (the rotational force applied to an object with a fixed pivot point, like swinging open a door or turning a bolt). The formula for torque is $\tau = F * d * \sin \theta$, where τ (tau) is the torque, F is the force being applied, d is the distance vector from the pivot point to the point where the force is applied, and θ (theta) is the angle at which the force is applied in relation to the vector d. We can't use Greek letters in MATLAB, and capital letters normally indicate matrices, so we'll restate the formula as tau = f * d * sin(theta).

Torque is often given in Newton·meters, but our hapless lab assistant has provided the distance value 8.2021 in feet. We know one foot equals 0.3048 meters so we can do the conversion. We were also expecting the θ value in degrees, but he provided the measurement 0.7854 in radians. The force applied was 15 N. (A Newton being 1 kg*m/s^2 or a force similar to that exerted on your hand when you pick up a large apple.)

Using MATLAB makes it simple to do the conversions, and as usual there is a built-in function to convert radians to degrees. The key point is that while it's easy to convert the units, you MUST keep track of them to make sense of the calculation. Our commands without comments used below do not help you keep track of the units well at all.

Enter the following commands in the Command Window.

```
>> f = 15;
>> ft = 8.2021;
>> rads = 0.7854;
>> d = ft * 0.3048;
>> theta = rad2deg(rads);
>> tau = f * d * sind(theta)
```

Notice that we don't put a semi-colon at the end of the tau equation, even though MATLAB suggests that we do. That would prevent the answer from displaying in the Command Window, though it would appear in the Workspace.

Creating and Editing a Script

It would be tedious to convert feet to meters, convert radians to degrees, and retype the torque formula for each piece of data, so we can turn those last three commands into a script. Once that's done, all we have to do is enter the *f*, *ft*, and *rads* values and then run the script. To turn commands into a script, try this simple method:

Press: ↑ *to show the Command History*

*Press and hold: **[Shift] key and select the last three commands** (the two conversion commands and the torque formula)*

*Right-click: **the selection in the Command History** and choose **Create Script** from the menu as shown in Figure 6.1*

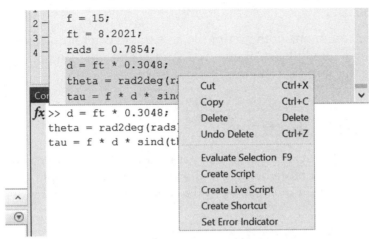

Figure 6.1   **Command History and Context Menu**

A tab opens in the Editor with those commands added to the file. The file is unnamed and unsaved. Add the disp command to the end of the file:

disp ('in Newton-meters')

 On your own, save the script to the current working directory naming the script Torque.

Note that the command is case-sensitive; you can't launch *Torque* by typing *torque*. When you launch the script by typing its name in the Command Window, each command in the script is executed in order, using the values of *f*, *ft*, and *rads* in the Workspace. If you do not have values for these variable set, you will see an error. Enter the following commands in the Command Window to test the script:

>> f = 12.5;

>> ft = 4.0056;

>> rads = 0.5236;

>> Torque

Your output should look like this:

```
tau =

    7.6307

in Newton-meters
```

There are three more ways to create a script:

- Choose Script from the New menu on the ribbon (requires a manual save to name the file)
- Hit the [Ctrl]+[n] key combination from the keyboard (requires a manual save to name the file)
- Type edit *scriptName* at the Command Window prompt (creates a named file)

Be aware that if your script includes the name of an existing Workspace variable, as ours does, that variable will be overwritten with the value stated in the script.

Comments in Scripts

You can include comments in the script file, to explain what it does and why. This is often a good idea. Comments do not echo to the Command Window during the script's execution, so they are only seen in the Editor.

Making a Script Interactive

A script can prompt the caller for input by using the input function, which can make a script more flexible and interactive than a simple list of commands. Add the following lines to the top of the *Torque* script in the Editor window:

f = input ('Enter force in Newtons: ');

ft = input ('Enter distance from the axis in feet: ');

rads = input ('Enter angle of force in radians: ');

```
Editor - C:\Documents\MATLAB\Torque.m
Torque.m
1    f = input('Enter force in Newtons: ');
2    ft = input('Enter distance from the axis in feet: ');
3    rads = input('Enter angle of force in radians: ');
4    d = ft * 0.3048;
5    theta = rad2deg(rads);
6    tau = f * d * sind(theta)
7    disp('in Newton-meters')
```

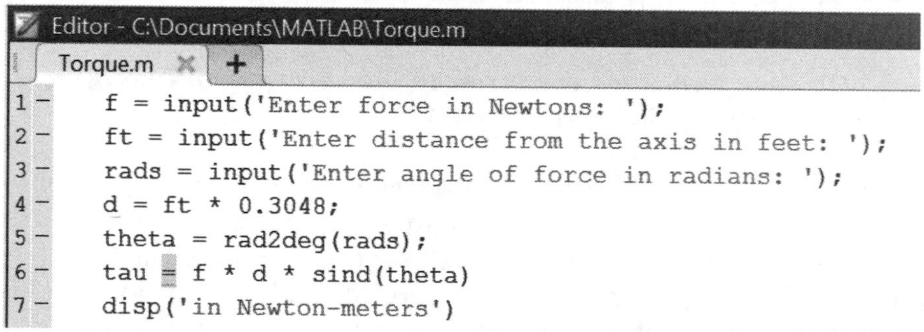

Figure 6.2

On your own, use the Command Window to run the Torque script.

Enter the values at the prompts (f = 15, ft = 8.2021, rads = 0.7854), press [Enter] after each number.

Your output should now look like Figure 6.3:

```
Command Window
>> Torque
Enter force in Newtons: 15
Enter distance from the axis in feet: 8.2021
Enter angle of force in radians: 0.7854
tau =
    26.5166
in Newton-meters
fx >>
```

Figure 6.3

There are a lot of approaches to the torque problem, but this shows one way to save you some typing and make your calculations more efficient.

Functions

A *function* can be thought of as a block of code that answers a specific question (such as "what is the square root of this value?") or perform a specific task (such as "display a plot of this equation"). In general, a full program is built of many functions, each of which performs a focused subtask. This modular design has several benefits:

- It's simpler to test individual, smaller components

- Knowing that each component works as expected lets you build complex programs with greater confidence

- Updating a program is safer because you can replace only the code that needs updating while leaving the rest of the known and well-tested code unchanged

- You can reuse code. If you've written a function that performs a task, you can use that function in any program you write to perform that same task

Like a script, a function contains a series of commands, is stored in a *.m* file, and can be called from the Command Window. Unlike a script, **a function can accept input parameters and return one or more output values**. Functions also use their own workspace instead of the Command Window's base workspace.

Built-In Functions

You've used dozens of MATLAB's built-in functions already: sum, length, logical, int32, ceil, mod, class, strcmp, ismatrix, islogical, magic, randi, struct2cell, rng and many others. Most of these accept an input value, such as ismatrix(M), and return a result, such as a logical 1 or 0, a new matrix, or a calculated value. Some functions, such as rng, accept an input value and perform a task (in this case setting the random number generator seed) but don't return anything. The built-in functions exist because they perform common tasks that MATLAB users are likely to need. It would be unreasonable and impractical to expect every user to write their own version of every function.

Note: For the record, the word "program" can refer to something like Microsoft Excel, Apple's iOS, a phone app, a web app, the software in a car's processor that controls fuel flow, or the code that makes a wireless router work. It's a general term for anything in any device that originates as code. Also, "programming" and "coding" are the same thing, as are "programmers" and "developers."

Note: *Most programming languages provide a large set of built-in functions.*

Of importance here is that the built-in functions are written using the same code elements that your own functions will use. In fact, you can see the source code for some functions that come with MATLAB. This can be a great learning tool. The which command, such as which *blkdiag*, gives you the path to a function's *.m* file, which you can open and read like any other *.m* file.

Custom Functions

Writing your own functions is the heart of procedural programming. You are unlikely to solve complex engineering problems or analyze huge data sets by keying commands into the Command Window or setting up scripts. Big problems need powerful and flexible data manipulations. A game app on your phone can be made up of hundreds of thousands, if not millions, of lines of code, organized in hundreds of functions, each of which performs its granular task and all of which interact and must work correctly. We're not going to do anything that extravagant, but this is where you start to do true programming.

You were introduced to creating your own functions in the tutorial on looping and conditional statements. Please take a few minutes to go back to that tutorial and review *Using the MATLAB Editor*. We'll proceed under the assumption that you have that information.

Function File Format

A MATLAB function file begins with the *function definition*, which looks like this:

```
function [retVar1, retVar2, ...] = functionName (param1, param2, ...)
```

Breaking it down:

The function keyword, which lets MATLAB know that this is a function file instead of a script.

Output variable(s): Either a single variable (brackets optional) or a bracketed list of comma-separated variables such as *[retVar1, retVar2]* in the code example above. The output variables are calculated and returned by the function. Output variables are optional. If there are no output variables, omit the equal sign as well.

Function name: The first function in the file must have the same name as the file (*functionName* in the code example above). Function names follow the same rules as variable names: any combination of a-z, A-Z, 0-9, and the underscore character, starting with a letter.

Input parameters: a comma-separated list of parameter names enclosed in parentheses such as *(param1, param2)* in the code example. When the function is called, these values must be included in the call. If there are no input parameters, the parentheses can be optionally omitted.

Here are some examples of function definitions:

```
function toast()

function toast
```

```
function toe = tic_tac(first_square)

function [x, y, z] = angry_cat(meow, hiss)
```

Fundamental function facts:

- The name of the .*m* file must also be the name of the first function in the file.

- Only the first function in the file can be called from the Command Window or from a function or script in another file.

- The .*m* file can contain as many functions as needed.

- Other functions in a file (known as *local functions*) can only be called by the first function or by each other.

- A function may receive input parameters ( ) and may output returned values [ ], but it does not have to have either input or output parameters.

Note: *While functions are common in programming languages, function definition syntax is language-specific. MATLAB's is a bit unusual.*

A function is closed with the keyword **end**, on a line by itself, as you closed a loop or a conditional branch. If you have only one function in the file, **end** is optional, but it's a good practice to include it.

```
function [retVar1, retVar2, ...] = functionName (param1, param2, ...)

      % Your code goes here

end
```

From Algorithm to Code

When given a problem to solve, programmers generally begin with an *algorithm*, which is a set of steps to solve the problem, written in plain text or *pseudo-code* (text that bears some resemblance to code but doesn't worry about proper syntax). This helps you conceptualize the problem without thinking about programming rules.

As an example, suppose you're asked to write a function to find the largest number in a matrix. (MATLAB has a **max** function that would make this easy, but assume you don't know that.) Here's a quick algorithm you might jot down for the problem:

1. Start with the first column in the first row

2. Automatically regard that value as the largest number so far

3. Move to the second column of the first row. Is that number larger than the first column? If so, that's the new largest number. If not, move on.

4. Keep looking at each column and comparing numbers until you reach the end of the row.

5. Go to the next row. Do it all again until you reach end of the last row.

OK, not bad, but it glides over a few things. Perfectly acceptable for a first pass. Next, we'll go back and add more details.

1. Create a variable *biggie* to hold the maximum value
2. *biggie* = the value at (1,1)
3. Look at the next column and compare its value to *biggie*
4. If the value is larger, set *biggie* to that value
5. If the value is equal or smaller, do nothing
6. Have we read the last column?
7. No: repeat steps 3-6
8. Yes: go to the next row
9. Repeat steps 3-6 for the next row, starting at column 1
10. Have we read the last row?
11. No: repeat steps 9-10
12. Yes: we're done
13. Report *biggie* to the caller

Note: *Another approach to problem solving is the use of a flowchart, as you saw in the if-elseif-else example in the loops and conditionals tutorial. If your brain likes a more visual approach, this might be for you. A quick web search will provide you with an endless list of tutorials.*

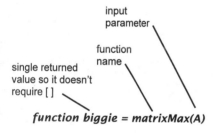

input parameter

function name

single returned value so it doesn't require [ ]

function biggie = matrixMax(A)

Tip: *Passing values in parameters can be confusing at first. Remember that a variable is just an alias for a value. The parameter A in the example is the name the matrix uses in the function. When you call the function, it doesn't matter what the matrix variable name is, just that it's a matrix.*

Some questions remain: How do you know how many rows and columns there are? What will you call the function? How can the function be flexible enough to accept a matrix of any size? But we probably have enough of a map to start translating the algorithm into code. Keep in mind that there is no "right" algorithm. Writing an algorithm is a bit of a craft, and, as with all crafts, the more you do it, the better you'll get at it. A little time honing an algorithm can save a lot of time coding.

First, we need to name the function so that we can create the file. Let's call it *matrixMax*. At the Command Window prompt, type this command:

> **>> edit matrixMax**

The editor opens a new file named *matrixMax.m* to hold the function code. You might find it helpful to add your algorithm to the file as a series of comments, translating each step to code underneath it.

The first line in the file must be the function definition, so we need to know what the function accepts as input and what it returns as output. The input for our function will be a matrix named A to search through and the output will be a single scalar that is the largest value in the matrix. In the Editor, type this line for the function definition:

> **function biggie = matrixMax(A)**

MATLAB immediately provides feedback. The input parameter, matrix variable *A,* is highlighted and the return value, variable name *biggie,* is underlined. When you hover your cursor over those variables you will see the information that *biggie* is not assigned a value to return and that *A* is unused. That's all true, but only because we haven't added any code to the function yet, so you can ignore it for now. Keep an eye out for that sort of warning as you enter more code, though. It can be useful information.

Next, we'll start creating the body of the function, indenting for clarity. First, keeping the algorithm in mind, we will *initialize* the return variable by assigning it the value of the first element in the matrix. Also include a comment that this will be our variables section.

Add these next lines as the beginning of the body of the function.

% Variables

biggie = A(1,1);

Notice that the equal sign was highlighted until you added the semi-colon. Most lines of code in a function end in a semi-colon to suppress the line's output. You don't want to see each line being executed. MATLAB highlights the equal sign to tell you exactly that. The semi-colons are optional, though. As an experiment, you can leave a few out, just to experience the noisy alternative when the function executes.

Create two more variables: the row count and the column count. The built-in **size** function returns both of those values as a 2-element vector. Separate them by using the **size** function's dimension parameter (1 for row, 2 for column). The function so far:

```
function biggie = matrixMax(A)

    % Variables
    biggie = A(1,1);
```

Add the next lines of code:

rows = size(A,1);

columns = size(A,2);

Look at the tab with the function name *matrixMax.m*. Notice the asterisk (*) next to the name (also in the title bar). That tells you that unsaved changes to the file have been made. MATLAB will not let you run an unsaved file.

Click back in the Command Window to make MATLAB automatically save the file for you.

The asterisk goes away. You can also click the Save button on the ribbon.

The next step in the algorithm is to start stepping through the matrix. To do this, we'll use a common solution: nested *for* loops. Notice the indentation to show which blocks of code are contained by each loop.

```
function biggie = matrixMax(A)

    % Variables
    biggie = A(1,1);
    rows = size(A,1);
    columns = size(A,2);
```

Add the next lines of code:

% Walk the matrix

for r = 1:rows

  for c = 1:columns

   end

end

Tip: *Add the* end *statements before you add the loop bodies, to make sure that each block has its corresponding* end, *which can get lost in larger programs, making a function invalid.*

The inner for loop is where we'll ask the question "Is this new value the largest yet?" That's a job for a conditional statement. We'll also append a comment to show you how a comment can be used in-line. The conditional statement will also need an end statement.

```
function biggie = matrixMax(A)
    % Variables
    biggie = A(1,1);
    rows = size(A,1);
    columns = size(A,2);

    % Walk the matrix
    for r = 1:rows
        for c = 1:columns
            if A(r,c) > biggie  % Is A(r,c) the largest yet?
                biggie = A(r,c);
            end
        end
    end
```

Lastly, add the *end* statement to close the function.

```
function biggie = matrixMax(A)
    % Variables
    biggie = A(1,1);
    rows = size(A,1);
    columns = size(A,2);

    % Walk the matrix
    for r = 1:rows
        for c = 1:columns
            if A(r,c) > biggie  % Is A(r,c) the largest yet?
                biggie = A(r,c);
            end
        end
    end
end
```

To test the function, drop back to the Command Window and use the next steps to create a 5 x 7 test matrix of random integers between 1 and 15. You will then call the function, passing this test matrix as the input parameter.

>> *T = randi(15,5,7);*

>> *matrixMax(T)*

And the answer returned by the function is 15. There is possibly more than one element in your matrix that is assigned the value 15, but we weren't looking for that information, just the highest value regardless of how many times it occurs. Test the function with some other matrices on your own.

Do you want to observe how the nested loops work? It's easy to demonstrate by inserting an **fprintf** line that outputs those values while the function runs. Enter the **fprintf** line as shown, and run matrixMax:

```
% Walk the matrix
for r = 1:rows
    for c = 1:columns
        fprintf('r = %d, c = %d \n', r, c);
        if A(r,c) > biggie  % Is A(r,c) the largest yet?
            biggie = A(r,c);
        end
    end
end
```

If you were walking a higher dimensional array, you would just nest one more loop for each additional dimension.

Return Values

Returning a single value from a function as we did above is a simple matter, but sometimes your function might need to return multiple values. Programming languages do this in different ways, and MATLAB's is one of the easiest.

Specify multiple return values as a comma-separated, bracketed list in the function definition. *Note that despite the brackets, this is not an array and the return values can be of different types.*

```
function [retVar1, retVar2, ...] = functionName(...)
```

For the caller to receive the return values, use the same format: a set of variables in brackets on the left side of the equal sign. For example, here is a function definition that returns a double x, a string y, and a matrix Z:

```
function [x, y, Z] = manyRetVals()
```

Again, the brackets do not make this an array; the variables are separate entities. The call to that function would look like the following. Commas are optional.

```
[a, b, C] = manyRetVals();
```

You aren't required to receive all the return values. If you supply fewer

Input parameter (our function only had one parameter and any variable, value, or expression can be passed in as that input)

Function call

matrixMax(T)

Tip: *The inputs and outputs of the function are only kept track of by their relative positions in the lists inside the parentheses ( ), or the brackets [ ]. Think of all the built-in functions you have used in MATLAB. They work this same way. For example: sin(x) where x is the input parameter you pass to the function.*

variables than the function returns, the output variables are received in order. For instance, this call receives only values x and y; value Z is lost:

> [a, b] = manyRetVals();

And this call receives only value x, while values y and Z disappear into space:

> [a] = manyRetVals();

Let's work a concrete example. We'll write a function that accepts a single matrix and returns a set of information about that matrix. This time we'll create the function from the ribbon.

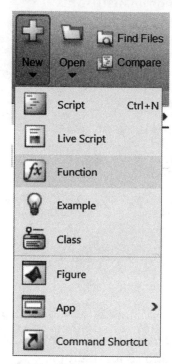

Figure 6.4

This method of function creation gives you a framework to fill in.

```
1    function [ output_args ] = untitled3( input_args )
2    %UNTITLED3 Summary of this function goes here
3    %     Detailed explanation goes here
4
5
6    end
7
```

Figure 6.5

On your own, replace output_args *with the variables* dimensions, rows, columns, maxval.

```
Editor - Untitled3*

Untitled3*  ×  +

1    function [dimensions, rows, columns, maxval] = untitled3( input_args )
2    %UNTITLED3 Summary of this function goes here
3    %    Detailed explanation goes here
4
5
6    end
7
```

Figure 6.6

Replace untitled3 *with the function name* matrix_facts.

```
Editor - Untitled3*

Untitled3*  ×  +

1    function [dimensions, rows, columns, maxval] = matrix_facts( input_args )
2    %UNTITLED3 Summary of this function goes here
3    %    Detailed explanation goes here
4
5
6    end
7
```

Figure 6.7

Save the file by clicking the Save button on the ribbon.

Name the file matrix_facts, *because you remember that the name of the file has to match the name of its first function.*

Figure 6.8

Replace input_args *with the variable* A.

```
Editor - Untitled3*

Untitled3*  ×  +

1    function [dimensions, rows, columns, maxval] = matrix_facts(A)
2    %UNTITLED3 Summary of this function goes here
3    %    Detailed explanation goes here
4
5
6    end
7
```

Figure 6.9

The included comments suggest that you should add some commentary to explain the purpose of the function. That's a good practice. Do so.

```matlab
Editor - C:\Documents\MATLAB\matrix_facts.m
matrix_facts.m  ✕  +
1   function [dimensions, rows, columns, maxval] = matrix_facts(A)
2   % Accepts a matrix and returns
3   %    - number of dimensions
4   %    - number of rows
5   %    - number of columns
6   %    - highest value in the matrix
7
8   |
9   end
```

Figure 6.10

Now for the function body. Add the lines as shown in Figure 6.11:

```matlab
Editor - C:\Documents\MATLAB\matrix_facts.m
matrix_facts.m  ✕  +
1   function [dimensions, rows, columns, maxval] = matrix_facts(A)
2   % Accepts a matrix and returns
3   %    - number of dimensions
4   %    - number of rows
5   %    - number of columns
6   %    - highest value in the matrix
7
8       dimensions = ndims(A);
9       rows = size(A,1);
10      columns = size(A,2);
11  end
```

Figure 6.11

If only there was some way to compute the maximum value. Hey, didn't we already write a function to do that? We can call our function from this one! Add the statement calling *matrixMax* as shown in Figure 6.12.

```matlab
Editor - C:\Documents\MATLAB\matrix_facts.m
matrix_facts.m  ✕   matrixMax.m  ✕  +
1   function [dimensions, rows, columns, maxval] = matrix_facts(A)
2   % Accepts a matrix and returns
3   %    - number of dimensions
4   %    - number of rows
5   %    - number of columns
6   %    - highest value in the matrix
7
8       dimensions = ndims(A);
9       rows = size(A,1);
10      columns = size(A,2);
11      maxval = matrixMax(A);
12  end
```

Figure 6.12

Click in the Command Window and key in the next series of commands. The first command clears all the existing variables in the Workspace.

The second command creates a test matrix. The third command calls our function and assigns its four return values to four local variables.

```
clear
A = randi(15, 5, 7);
[dim, r, c, mx] = matrix_facts(A)
```

You should see a result similar to this (*mx* might differ):

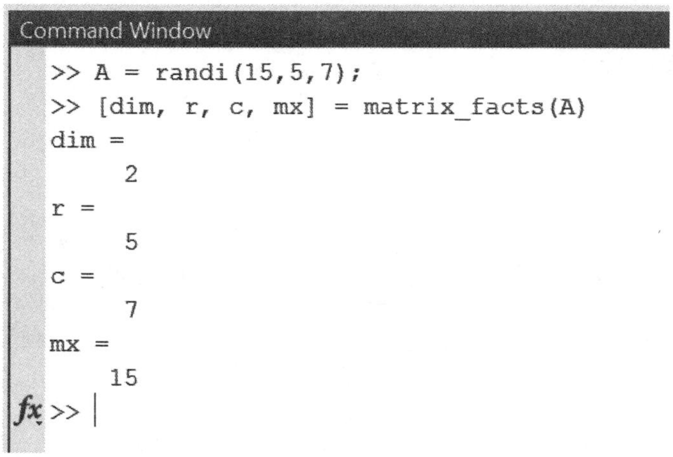

```
Command Window
>> A = randi(15,5,7);
>> [dim, r, c, mx] = matrix_facts(A)
dim =
     2
r =
     5
c =
     7
mx =
     15
fx >> |
```

Figure 6.13

The Workspace now contains the new variables, with the values returned from the function.

Local Functions

In MATLAB, a *local function* is a function that can only be called from within its file, either from another function in the same file or as part of a script. It cannot be called from any other file or from the Command Window. A file can contain as many local functions as necessary.

Let's revisit the program we wrote in the Return Values section, but instead of calling the external *matrixMax* function, we'll write a local function to find that value.

Because the first function in a file must match the name of the file, any local function must be added *after* the main function.

In the matrix_facts.m file, change the line,

    maxval = matrixMax(A);

to this:

  maxval = myMax(A);

To the end of the file, after the *matrix_facts* function's **end** statement, add the code shown in Figure 6.14 for the *myMax* local function:

```matlab
Editor - C:\Documents\MATLAB\matrix_facts.m

matrix_facts.m   ✕   +

1    function [dimensions, rows, columns, maxval] = matrix_facts(A)
2    % Accepts a matrix and returns
3        %    - number of dimensions
4        %    - number of rows
5        %    - number of columns
6        %    - highest value in the matrix
7
8        dimensions = ndims(A);
9        rows = size(A,1);
10       columns = size(A,2);
11       maxval = myMax(A);
12   end
13
14   function max_value = myMax(M)
15   % Finds the largest value in a matrix
16
17       v = max(M);
18       max_value = max(v);
19   end
```

Figure 6.14

The *max_value* function calls the built-in **max** function. When applied to an *m* x *n* matrix, the **max** function returns a vector of length *n* that contains the maximum value in each column. In turn, calling the **max** function on that vector returns its maximum value, which is the maximum value in the matrix. If you want to get fancy, you can reduce those two function calls to one:

max_value = max(M(:));

The colon operator by itself converts the entire matrix M to a single column vector, which the **max** function can work on directly.

A Variable's Scope

The concept of variable scope is extraordinarily important. *Scope* refers to the set of statements that can see and use a variable. A variable can be of *local*, *persistent*, or *global* scope.

Recall that a variable is essentially an alias for an address in memory. When a value is passed to or returned from a function, the value itself is passed, not the memory address, and that value is put into a memory address accessible by that function. The total address space that a set of commands can access is its scope.

- A function can't see the address space of the Command Window.

- The Command Window can't see the address space of a function.

- Functions can't see the address space of other functions.

- Scripts use the address space of the Command Window, but a local function in a script can only see its own address space.

Local Scope

To demonstrate local scope we'll use two variables with the same name.

First, clear the Command Window and its Workspace using the clc ***and*** clear ***commands.***

Next, define the new local variable *local_var*, plus a few other local variables. using the following statements in the Command Window:

>> *local_var = 1000;*

>> *a = logical(1);*

>> *b = 57;*

>> *c = 'autumn leaves';*

>> *X = magic(4);*

On your own, create a new function called scope ***that accepts no parameters and returns no value.***

Add lines 2 and 3 before the end statement as shown in Figure 6.15.

```
Editor - C:\Documents\MATLAB\scope.m
  scope.m  ✕  +
1    ⊟ function scope()
2  –       local_var = 20;
3  –       local_var = local_var + 10;
4  –   └ end
```

Figure 6.15

We're going to get ahead of ourselves a bit with a debugging technique called a *breakpoint*. A breakpoint stops a function at that line and allows you to manually control the rest of the execution. In the Editor, click the dash next to the line number 2 as shown in Figure 6.16. A circle replaces the dash to show you the breakpoint. (The circle is gray if the file hasn't been saved yet, and red otherwise.)

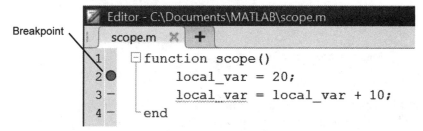

Breakpoint

```
Editor - C:\Documents\MATLAB\scope.m
  scope.m  ✕  +
1    ⊟ function scope()
2 ●       local_var = 20;
3  –       local_var = local_var + 10;
4  –   └ end
```

Figure 6.16

When you launch the *scope* function, execution will stop right before it executes the line with the breakpoint.

But before you launch the function, take a good look at the Workspace. The variables listed there are the local variables available to the Command Window and scripts. This is the *base workspace*.

Particularly note that the value of *local_var* is 1000 (see Figure 6.17).

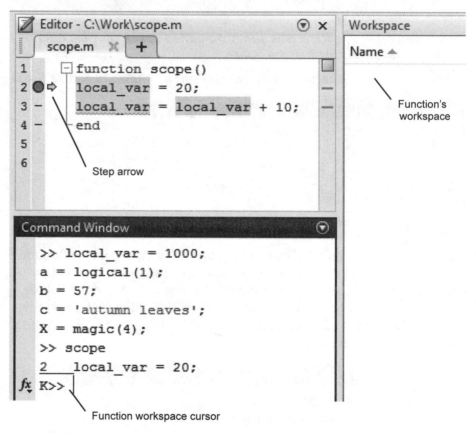

Figure 6.17   **Base Workspace Variables**

On your own, launch the function by issuing the command scope.

The breakpoint does its job and you're sitting at a stoplight at the beginning of the scope function. A green arrow shows you where you are.

Note: *Methods exist in some languages for passing memory addresses instead of values, normally involving a variable type known as a pointer. Pointers are used a lot in C and C++, but aren't widely supported in MATLAB.*

Figure 6.18

Note: *In many languages, a variable defined inside a loop, including the loop counter, exists only inside that loop and goes out of scope once the loop exits. This is not true in MATLAB.*

Notice that the Workspace is empty. This is the *function's* Workspace – all those variables in the base Workspace aren't available here.

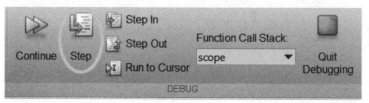

Figure 6.19

*On the ribbon, from the Debug section, click the **Step** button*

This advances one line in the function, executing the line `local_var` = `20` and adding that variable to the function's Workspace.

Click the Step button once more, executing the next line in the function, adding 10 to the local_var value.

Notice that the value in the Workspace has changed, as you'd expect.

Click the Continue button to finish executing the function and return to the Command Window.

Notice that all the original Workspace variables are back and that *local_var* again equals 1000.

That's scope in action. The *local_var* variable in the function workspace is not the same *local_var* variable in the base workspace, despite having the same name.

Global Scope

You can override local scope by declaring a variable as *global*, which means that it can be shared across all functions, scripts, and the base workspace. We're not going to say much about global variables because their use is generally discouraged. There are scenarios in which a global variable is the legitimate choice, but more often they're a sign of poor program design and can lead to some very confusing errors.

To define a global variable, simply precede it with the keyword *global*.

```
global b;
```

Unlike a normal variable, you cannot declare a global variable and give it a value on a single line. The two operations must be separate.

```
global b;

b = randi(2000, 13, 22);
```

To access a global variable, you again need a declaration line to bring the variable into the current Workspace before you can use it in a statement. For instance, in a function you might have these lines:

```
global b;    % Bring the variable into the function

var1 = b;    % Use the variable

b = var2;    % Set the variable
```

When you change the value of global variable *b*, it affects that variable's value wherever it is used, in any other function or in the Command Window. In other words, its scope is global.

Recursion

A *recursive* function is a function that calls itself. To avoid an endless spiral of calls, a recursive function must contain a stop condition. Once the stop condition is met, the stack of function calls reverses direction and returns its value to its caller until the original caller is reached with the final result.

The classic illustration of recursion is the calculation of a factorial. The factorial of a number *n* is notated as *n!* and defined as shown here, with 1! defined as equal to 1.

```
1 * 2 * 3 * … * n
```

Take 4! as an example.

```
1 * 2 * 3 * 4
```

Multiplication is associative, so that formula can be rewritten in any of these ways:

```
(1 * 2 * 3 ) * 4
((1 * 2) * 3 ) * 4
(((1) * 2) * 3 ) * 4
```

Which shows us the following:

```
4! = 3! * 4
3! = 2! * 3
2! = 1! * 2
```

Those formulas all use the same pattern of `x! = (x-1)! * x`, so we can design a function around that general formula. Here's an algorithm:

1. Begin with a number n

2. If n = 1, then n! = 1 and we're done

3. If n ~= 1, then get (n - 1)!

4. Multiply (n - 1)! times n

Our recursive stop condition, therefore, is n = 1.

Here is the recursion function. Take a couple minutes to really look at this to understand what's going on. Follow it through, jotting down the values that are being passed and returned.

```
Editor - C:\Documents\MATLAB\recurse.m
recurse.m  ✕  +
1    function fact = recurse(n)
2        if n == 1
3            fact = 1;
4        else
5            fact = n * recurse(n-1);
6        end
7    end
```

Figure 6.20

Persistent Variables

Normally, when a function exits, its local variables disappear forever. However, by declaring a variable *persistent*, its value is held in memory so that the next time that function is called, the variable's last value is still there (as long as you haven't restarted MATLAB).

The following function *fthw* (which stands for "factorials the hard way") shows how a persistent variable works.

```
Editor - C:\Documents\MATLAB\fthw.m
fthw.m  ✕  +
1    function f = fthw(x)
2        persistent y;
3        if x == 1
4            y = 1;
5        else
6            y = y * x;
7        end
8        f = y;
9    end
```

Figure 6.21

Calling that function from the Command Window starting with x = 1 and increasing x by 1 with each subsequent call gives you a manual factorial with no recursion.

```
Command Window
>> fthw(1)
ans =
     1
>> fthw(2)
ans =
     2
>> fthw(3)
ans =
     6
>> fthw(4)
ans =
    24
fx >>
```

Figure 6.22

Why would you do it that way? You wouldn't; this is for demonstration purposes only. Like global variables, persistent variables have their place, but it isn't common to find yourself in need of one.

You have finished Tutorial 6.

Key Terms

algorithm	*initialize*	*persistent variable*	*scripts*
function definition	*input value*	*pseudo-code*	
functions	*local functions*	*return value*	
global scope	*local scope*	*scope*	

Key Commands

%	edit	function	size
disp	end	input	which

Exercises

 Exercise 6.1

Rewrite the terminal velocity script, first seen in the Introduction and last modified in the loops and conditionals tutorial, as a function which accepts a planet name (as a character array string) as its input parameter and returns the terminal velocity for that planet. Hint: To avoid case-sensitivity issues, you can use the **lower** or **upper** function to convert the input string into a standard form that your function expects.

Exercise 6.2

Modify the *matrixMax* function defined in the *From Algorithm to Code* section so that it also returns the row and column of the *last* occurrence of the maximum value, as a 1 x 2 row vector.

Exercise 6.3

Write a function called *torque1* to calculate torque, which was discussed in the Scripts section. The function will accept five parameters:

- a force *f*, in Newtons
- a distance *d* from the pivot point
- a unit indicator for the distance value: 'f' for feet or 'm' for meters
- an angle at which the force is being applied
- a unit indicator for the angle value: 'r' for radians or 'd' for degrees

The function will return the value calculated by the formula *tau = f * d * sin(theta)*. Convert distance if necessary and select the proper sine function based on its unit indicator.

Exercise 6.4

Write a function called *torque2* to calculate torque, based on the code in Exercise 6.3. This version accepts three parameters:

- a three-column matrix, where column 1 contains the force value, column 2 contains the distance value, and column 3 contains the force angle value
- a unit indicator for the distance value: 'f' for feet or 'm' for meters
- a unit indicator for the angle value: 'r' for radians or 'd' for degrees

The function will return a column vector with each row being the torque value for the corresponding row of the input matrix. As in Exercise 6.3, the torque value is calculated by the formula *tau = f * d * sin(theta),* and the unit indicators are used to determine whether a distance conversion is necessary and to select the proper sine function.

 Exercise 6.5

Write a recursive function called *reversed* that accepts a character array string and reverses its order. Do not use the built-in **reverse** function. Hint: Your solution can use the built-in **extractBetween** function. Pay close attention to the data types that you're passing and receiving.

DEBUGGING AND ERROR HANDLING

Introduction

Despite a well-designed algorithm and careful coding, many scripts and functions don't work on the first try. They may crash with an error, return an unexpected answer, or get stuck in an infinite loop and not return at all. The more complex the function, the more likely it is to contain a logical error or an unexpected data flow. To find out where it went wrong, you must *debug* the code. MATLAB provides a *debugger* that enables you to manually control the execution of a function, watching it work from the inside.

Everyone with a computer, tablet, or phone has had the experience of a program that suddenly fails, often displaying an inscrutable message about what went wrong – a "bad user experience." No one can predict every possible failure condition, but a good programmer attempts to foresee the most likely trouble spots and includes code that gracefully handles (or even prevents) a problem. This is called *error handling*.

Debugging and error handling are a craft that you'll improve at with experience. In this tutorial, we'll present some basic techniques.

Syntactic Errors

The word *syntactic* means a matter of *syntax* – following the rules, such as not using an octothorpe (#) in a variable name. Syntactic errors are the easiest type to find and fix because MATLAB can recognize them as you enter them, and will tell you about it. For instance, enter the following in the Command Window:

```
>> clear
>> a * 5
```

MATLAB displays the error "Undefined function or variable 'a'." No variable named *a* exists because you cleared the Workspace, and you can't use a variable before it has a value. Now enter this command, exactly as given:

```
>> cos(sin(0.45)
```

There's a parenthesis missing on the end, so MATLAB displays the error "Expression or statement is incorrect--possibly unbalanced (, {, or [." It even points an arrow at where the missing parenthesis should go. It follows this with a "Did you mean" suggestion.

Objectives

When you have completed this tutorial, you will be able to

1. Know the difference between syntactic and semantic errors.

2. Recognize and fix syntactic errors based on MATLAB's visual feedback and suggestions.

3. Set a breakpoint to enter the debugger.

4. Manually execute a function using the Step and other buttons on the Debug toolbar, or their keyboard equivalents.

5. Find problems in your code using both the MATLAB debugging functionality and common manual practices.

6. Demonstrate good programming practices for input verification and standardization.

7. Use the try-catch construction to avoid program crashes.

```
Command Window
>> cos(sin(0.45)
 cos(sin(0.45)
                 ↑
Error: Expression or statement is incorrect--possibly unbalanced (, {, or [.

Did you mean:
fx >> cos(sin(0.45))|
```

Figure 7.1   **Matlab Error Message**

The suggestion is correct, so all you need to do is press [Enter]. MATLAB can't always extrapolate what you meant, but it often gets it right.

Now press the ↑ key on your keyboard to show the Command History. Notice that the command with the missing parenthesis has a red dash next to it. This is how MATLAB shows you that this command generated an error.

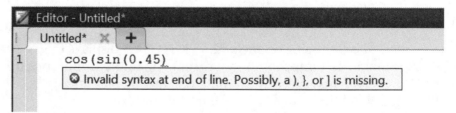

```
 — cos(sin(0.45)                                              ⌄
fx >> cos(sin(0.45)|
```

Figure 7.2   **Corrected Command and Error Indicated in History**

On your own, scroll to that line in the History

*Right-click: on that line and choose **Create Script**.*

In scripts and functions, MATLAB doesn't display an error message in-line as it does in the Command Window. Instead, it adds a small red squiggle under any spot where it sees a syntax error.

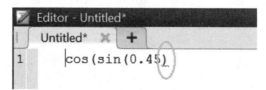

```
Editor - Untitled*
Untitled*  ✕  +
1    |cos(sin(0.45)
```

Figure 7.3   **Syntax Error in the Editor Window**

On your own, hover your cursor over the squiggle to see what MATLAB reports.

```
Editor - Untitled*
Untitled*  ✕  +
1    cos(sin(0.45)
     ⊗ Invalid syntax at end of line. Possibly, a ), }, or ] is missing.
```

Figure 7.4

Watch for these squiggles – they're small but powerful. Notice the small alley running down the far right side of the Editor window.

At the top of the alley the colored box shows:

- green if there are no errors,

- yellow if there are warnings or potential errors,
- red if there are definite errors.

Each line with an error or warning has a corresponding red or yellow dash in the alley. Hovering over one of those dashes displays its associated line number and error. This can be a big help in a long file.

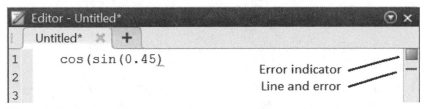

Figure 7.5

Tip: *Missing a closing parenthesis is probably the most common syntax error. MATLAB helps you avoid this by highlighting an open parenthesis when you type its matching closing parenthesis. If the opening parenthesis is off-screen, a small note with the opening parenthesis and its line number appears.*

If you didn't notice the error and ran this script, you'd see the same error message in the Command Window that you saw earlier, so MATLAB always finds a way to tell you about a known problem.

A syntax error causes the execution of a script or function to terminate immediately, so only the first error is reported in conjunction with that run. If there's a second syntax error, it will break the next run. However, with all the ways MATLAB brings this kind of error to your attention, you should be able to correct syntax errors before you ever execute the script or function.

Semantic Errors

A *semantic error* is an error in your code's logic. There's nothing syntactically wrong with the code, but it doesn't do what it's supposed to do. In some cases, this results in an error message from MATLAB (called a *runtime error*), but in some cases there's no error to report. Always remember – and this is terribly important – *a computer only does what a programmer tells it to do*. It's a machine – wires and transistors and electricity and dust. It can't second guess the intention of your logic. For instance, there's nothing to stop the following function from giving you incorrect information.

```
function x = addSeven(num)

    x = num + 7;

    x = 10;

end
```

If you call `addSeven(3)`, it will even return a correct answer! In all other cases, the answer is wrong, but the code is legitimate. MATLAB executes each line in turn and duly returns the last value of x, because that's what we told it to do.

Using the Debugger

We've provided a broken version of the terminal velocity function that we'll use to demonstrate debugging and error handling.

Open the data file terminalVelocityDebug.m in the MATLAB editor

First, run the function as provided. Call the function for the planet Mars using this entry in the Command Window:

>> ***terminalVelocityDebug('Mars')***

You will see this runtime error:

> Undefined function or variable 'dia'.
>
> Error in <u>terminalVelocityDebug</u> (<u>line 9</u>)
>
> area = pi*((dia/2)^2) % Cross-sectional area of the falling object,

Notice that the line number is underlined. This is a link that takes you to that line in that file – very helpful in a large program. According to the error, the variable *dia* in the equation on that line is *undefined*. But didn't we define that earlier? Oops, we spelled it wrong:

> <u>diam</u> = 2.5;   % Diameter of the sphere, in cm

MATLAB has tried to warn us about the problem by underlining the word "diam" with the red squiggly line. Hovering over the variable name, we're told that the value is never used in the code, which should have set our alarm bells ringing.

This function was originally a script, so maybe when we converted it we typed some of the lines and cut-and-pasted others, leading to this discrepancy in the *dia/diam* variable names. Cut-and-paste is a marvelous, time-saving feature, but it's a common source of errors, too. Because this variable is used only once, we can change the name in either the variable definition or in the equation, but we'll change the variable definition.

> ***On your own, edit the line of code as follows:***

> dia = 2.5;   % Diameter of the sphere, in cm

Notice that when you change the variable name, MATLAB highlights all other instances of that variable in the function, linking them. If you change the variable name again, a message appears offering to change that variable name in all the other places that it occurs – very helpful for avoiding a missed change.

Before you run the function again, notice that the equal sign in the next line is highlighted in yellow and underlined with the red squiggle.

> ***Hover your cursor over the equal sign***

A warning tip appears that there's no semi-colon at the end of that line.

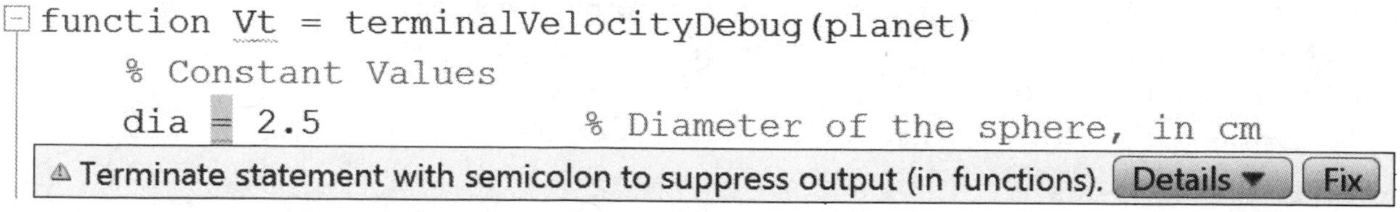

Tip: *Don't forget you can drag and drop variable names from the Workspace into the Editor and into the Command Window. This is a great time-saver, especially for long variable names and keeps them from being misspelled.*

```
function Vt = terminalVelocityDebug(planet)
    % Constant Values
    dia = 2.5                    % Diameter of the sphere, in cm
```
⚠ Terminate statement with semicolon to suppress output (in functions). [Details ▼] [Fix]

Figure 7.6

There are two buttons in the warning. Click the Details button to see a long-winded explanation of why it's good to use a semi-colon. Click the Fix button to automatically add the semi-colon. In some cases you don't want to suppress the output, but MATLAB always gives the warning. This is fine. Warnings can be safely ignored if you have a good reason. This isn't one of them, though, so:

On your own, add a semi-colon to that line

Run the function again with 'Mars' as the planet

Hmm. No errors, but no terminal velocity either. Try assigning the return value to a variable. Type the following statement in the Command Window:

>> v = terminalVelocityDebug('Mars')

Aha! Now we get an error: Output argument "Vt" (and maybe others) not assigned during call to "terminalVelocityDebug".

We definitely have a line that calculates that value, so why isn't it being assigned? Is it possible that the program flow never reaches the line? That could happen if the assignment statement was in a loop that was never executed or a branch of a conditional that's never reached. It could also happen if there is a **return** command earlier in the function. However, in this case, it's much simpler. Look at the *error/warning alley* at the right. It shows a yellow square (for potential problems) and two errors.

Hovering over the first of the error indicators, the message tells us in less definite terms what the error told us outright: that the output variable *Vt* isn't receiving a value before the function returns.

Hovering over the second error, we're told that the variable is assigned a value, but it's never used in the function. Compare the two errors:

⚠ Line 6: The function return value 'Vt' might be unset. [Details ▼]

⚠ Line 45: The value assigned to variable 'vt' might be unused. [Details ▼]

There's the problem – MATLAB is case-sensitive, so *Vt* and *vt* are two independent variables. We should have noticed the red squiggle underlines that both the return variable and the variable in the assignment statement have.

On your own, change the variable *vt* in the assignment line at the end of the function to *Vt*

Run the function again

Success! The dropped sphere will reach a terminal velocity of v =

3.244722524809211e+04 cm/s when falling through the Martian atmosphere.

To be honest, all the errors that we've fixed so far are pointed out by MATLAB and easy to find and fix. In real life, you would have probably fixed them as you entered the code.

But are we done now? Is everything working? Maybe and maybe not. As you've seen, the function works fine for the input parameter 'Mars'.

On your own, try some of the other planets, which should also work (although Mercury returns an odd result).

But now enter the following statement into the Command Window:

>> *terminalVelocityDebug('Pluto')*

You'll see this error:

Undefined function or variable 'gravity'.

Error in terminalVelocityDebug (line 45)

Vt = sqrt((2 * mass * gravity) / (atmosphericDensity * C * area));

The terminalVelocityDebug function only works for *planets* in our solar system, not for dwarf planets. But what is the function doing with this invalid input to cause this error? To find out, we can walk through the function, watching over its shoulder as it executes.

Breakpoints

You were introduced to breakpoints in the Functions and Scripts chapter, but we'll review. A *breakpoint* pauses the execution of a function or script at a specific line and puts MATLAB into debugging mode. A breakpoint can be set on any executable line. Executable lines have a dash in the breakpoint alley next to their line numbers.

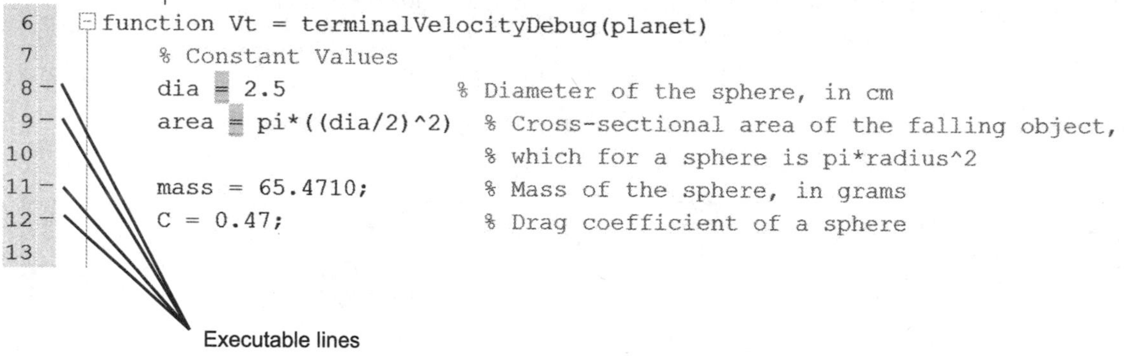

```
 6  ⊟ function Vt = terminalVelocityDebug(planet)
 7        % Constant Values
 8  –     dia = 2.5                % Diameter of the sphere, in cm
 9  –     area = pi*((dia/2)^2)    % Cross-sectional area of the falling object,
10                                 % which for a sphere is pi*radius^2
11  –     mass = 65.4710;          % Mass of the sphere, in grams
12  –     C = 0.47;                % Drag coefficient of a sphere
13
```

Executable lines

Figure 7.7

Clicking a dash adds a breakpoint, shown with a circle – gray if the file is unsaved or can't be run; red otherwise. You can add as many breakpoints as you need.

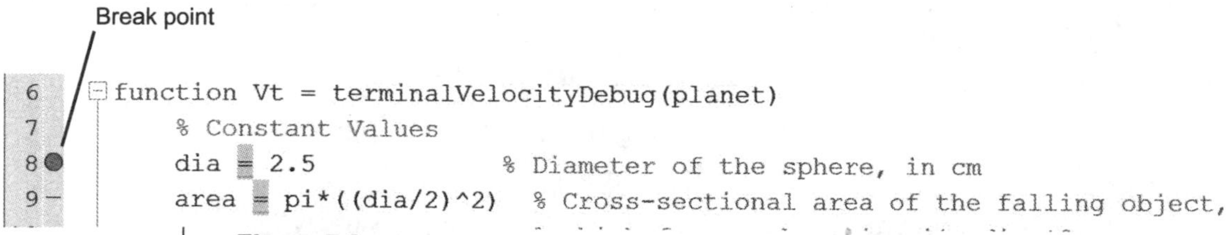

Break point

```
 6  ⊟ function Vt = terminalVelocityDebug(planet)
 7        % Constant Values
 8  ●     dia = 2.5                % Diameter of the sphere, in cm
 9  –     area = pi*((dia/2)^2)    % Cross-sectional area of the falling object,
```

Figure 7.8

When you run the function now, it executes until it encounters the first breakpoint and then pauses. At that point, several things happen:

- The Command Window displays the full line that the function is paused on (the line number is a link that will take you to that line in the Editor)

- The Command Window prompt changes from >> to K>>
- The Editor window adds a small green arrow pointing to the line that it's paused on

```
 6    ⊟ function Vt = terminalVelocityDebug(planet)
 7          % Constant Values
 8 ●⇨       dia = 2.5              % Diameter of the sphere, in cm
 9 ─        area = pi*((dia/2)^2)  % Cross-sectional area of the falling
10                                 % which for a sphere is pi*radius^2
```

Figure 7.9

- The Workspace shows the contents of the function's workspace instead of the base workspace
- The **Run** toolbar on the ribbon is replaced by the **Debug** toolbar

The Command Window in Debug Mode

Before we go into the details of the toolbar, a few words on the Command Window. When the function is paused, the Command Window prompt changes to K>> and its scope is the function's workspace. Other than that, the Command Window is unchanged. You can do anything you normally do in the Command Window – create variables, run equations, set new values for existing variables, check types and values, call functions wholly unrelated to the paused function, etc.

The Debug Toolbar

Here is the default Debug toolbar as shown on the ribbon when the function is paused.

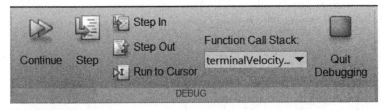

Figure 7.10

The commands and options in the toolbar allow you to manually control the function's execution. Some buttons have a keyboard shortcut as well.

- **Continue (F5)**: Executes the function normally until either the next breakpoint is encountered or the function exits.
- **Step (F10)**: Executes the current line and moves to the next line.
- **Step In (F11)**: If the line calls another function, opens and steps into that function, still debugging line by line. If you click **Step** instead of **Step In**, the secondary function is executed normally.

- **Step Out (Shift+F11):** When you are in a called function that you've stepped into, finishes executing that function, returns to the original function, and pauses on the line with the function call as though it had a breakpoint (although in this case the line has already been executed).

- **Run to Cursor:** Click anywhere in the file past the current point and click this button to execute the function up to that line and pause. If the cursor is not in an executable line, the function pauses at the first executable line after the cursor. If a breakpoint exists between the current position and the cursor, the function pauses at the breakpoint.

- **Function Call Stack:** Allows you to examine the workspace of any currently running functions or the base workspace.

- **Quit Debugging (Shift+F5):** Aborts the operation and returns to the Command Window and base workspace. The [Ctrl] + [C] keyboard shortcut has the same effect. This is particularly useful in escaping an infinite loop.

MATLAB provides a set of commands to use in place of the Debug toolbar. Most of these commands begin with "db" for "debug". For instance, typing dbquit at the Command Window prompt exits the debugger. The effect of the dbstep command is identical to clicking the Step button, while dbstep in and dbstep out duplicate Step In and Step Out. However, you can also use the dbstep command to execute a specified number of lines, which can't be done from the toolbar. These commands can be used in a script to automate debugging. We will only be using the toolbar, but if you want to learn more about these commands, see the Debugging topic in the MATLAB help.

OK, let's use the buttons on the toolbar to walk the function.

Set a breakpoint at line 8 (dia = 25)

Use the following entry to call the function from the Command Window with the parameter 'Jupiter'.

>> terminalVelocityDebug('Jupiter')

The function will pause at the breakpoint, with the green arrow pointing at that line.

```
6   function Vt = terminalVelocityDebug(planet)
7       % Constant Values
8       dia = 2.5              % Diameter of the sphere, in cm
9       area = pi*((dia/2)^2)  % Cross-sectional area of the falling
10                             % which for a sphere is pi*radius^2
```

Figure 7.11

Notice that in the Workspace window – which now shows the function's workspace – the only variable defined so far is the input variable *planet*.

Click: **Step button** to execute line 8

The variable *dia* is defined and joins *planet* in the Workspace.

Click: somewhere in **line 12 (C = 0.47)**

Click: **Run to Cursor button**

Lines 9-11 are executed, defining the area and mass variables.

Click: **Step button**

The variable C is now defined as well.

Because they are not executable lines, the comments that follow line 12 are skipped over and we're taken to the **switch** statement.

Click: **Step button**

We're now inside the **switch** statement at the first case: Mercury.

Click: **Step button** *once again*

This takes you to the next case: Venus. Do you remember the exercise in Loops and Conditionals that had you translate a **switch** statement into a series of **if-elseif** statements? By stepping through the **switch** statement, you can see it working through those options in that way, looking for a match with *planet*.

On your own, continue to hit Step until the green arrow points at line 34 (case 'Jupiter')

Now to enter the Jupiter case, since that equals the value of *planet*.

Click: **Step button** *to enter the Jupiter case*

On your own, continue to click Step

As you step through the code the *gravity* and *atmosphericDensity* variables are set. Then next click of the Step button moves the execution out of the switch statement altogether since there's nothing more to be done – the case was found and only one case can be executed.

All our variables are now defined and the line that calculates the terminal velocity is finally reached.

Once you're at the end of the function;

Click: **Continue** *to wrap it up and return to the Command Window. (You could also hit Step two more times with the same result.)*

Now that you've stepped through the function with a successful example, let's do it again, but with the error-causing value 'Pluto'. At the Command Prompt, enter:

>> terminalVelocityDebug('Pluto')

Once more, MATLAB pauses execution of the function at the breakpoint at line 8 (dia = 25). We don't need to step through setting all the initial variables again, so

Click: **the original breakpoint at line 8 to remove it**

Add a breakpoint at line 18, the switch **statement on your own**

Notice that you can add and remove breakpoints as needed during debugging, not just before you run the function.

Click: **Step to enter the switch statement**

Note: *In some programming languages, you must explicitly exit the switch statement as part of the case block. If you don't, program flow falls through to the next case, executes that as well, and continues to execute the following cases until it's told to exit or runs off the end. This is useful in executing code for a subset of cases.*

On your own, step through the statements

You'll find that you never enter a case, because there is no case for Pluto. When you reach the calculation, *gravity* and *atmosphericDensity* are not listed in the Workspace because you never entered a case to set them. If you were to click Step at the *Vt* calculation, MATLAB would display the error message and the function would exit. You know what the problem is, now so you don't need to continue. To exit quickly:

Click: **Quit Debugging** *button*

We'll fix this issue in the next section.

Error Handling

When we think about what can go wrong with the terminal velocity function and how we can we can deal with those potential problems, there are several improvements that we can make to the code. This proactive approach is a hallmark of good programming.

The weak spot in our function is the input parameter. MATLAB is *type-flexible*, so our input parameter will accept anything – a character array (as we expect), but also a number, matrix, structure, or any other type. We can't control what the function receives, but we can control how we handle it.

Standardizing and Verifying Input Data

Verifying input is a good programming practice. For instance, if your function expects a number, *x*, between 10 and 20, you could add this code at the top of the function:

```
if x < 10 || x > 20

    disp('Invalid input, must be between 10 and 20');

    return;  % No point in continuing, so exit the function

end
```

Perhaps your function expects a scalar. This code block would ensure that you get one:

```
if ~isscalar(x)

    disp('Invalid input, scalar value required');

    return;

end
```

The terminal velocity function can use only a limited set of strings. As it happens, the function works whether it receives a character array (single quotes) or a string (double quotes), but you could restrict the input to character arrays only.

```
if ~ischar(planet)

    disp('Invalid input, character array required');

    return;

end
```

When working with text input, *case-sensitivity* can cause a problem. Try running the next statement at the Command Prompt:

>> *terminalVelocityDebug('mars')*

Error! Error! MATLAB can't extrapolate that you meant 'Mars' instead of 'mars'. When case isn't a factor in the result, standardizing text input with the upper or lower functions neutralizes the effect of case-sensitivity.

Add the following statements to *terminalVelocityDebug.m,* ***immediately following the function definition in the Editor window:***

% Standardize input

planet = lower(planet);

On your own in the switch ***statements uncapitalize the planet names so that the strings are all lower case.***

The function will now work regardless of the casing of the input. (The lower function ignores any non-text input, so it won't fail on something like a matrix of numbers. To be extra cautious, you can put the *lower* statement inside a conditional block.

if ischar(planet) || isstring(planet)

% Standardize text input

planet = lower(planet);

end

Because case is unimportant to the result of *terminalVelocityDebug*, this is a better way to handle imprecise input than aborting the function.

To test that it works do this:

Set a breakpoint at the switch statement

Run the function at the Command Prompt with the following statement:

>> *terminalVelocityDebug('NepTunE')*

Paused at the breakpoint, notice the value of the variable *planet* in the Workspace: 'neptune'.

Click: **Continue** *button*

to complete the execution and you should see the result 11,858.

Let's get back to the original problem: the *gravity* and *atmosphericDensity* variables are unassigned if *planet* doesn't match any case in the switch statement. We have two approaches to solve that problem.

The first approach is to define and assign default values to those variables before the switch statement. If *planet* matches a case in the switch statement, the variables are assigned new values; if not, they retain the default values. Either way, the variables are defined when they're needed in the calculation, regardless of the program flow. In general, this is a good practice if those base values have a specific meaning. For instance, a counter would be initialized at 0. The defaults also could be absurd values that would demonstrate an obvious error,

Tip: *Don't forget that the format command determines how the result is displayed.*

such as a person's age being 999. For gravity and atmospheric density, -1 would be a suitable absurd value for both in this context, but we have a better solution.

A **switch** statement has the option of an **otherwise** block that is executed when no other case matched the input. Add the following **otherwise** block to the end of the **switch** statement.

```
    case 'Uranus'
        gravity = 887;
        atmosphericDensity = 0.00042;
    otherwise disp ('Invalid input, not a planet in our solar system')
        return;
end
```

Figure 7.12

This is the best solution because any case-standardized input other than the eight planet names will be handled by this block, which terminates the function before attempting the terminal velocity calculation.

Try/Catch Blocks

There is one last piece of bulletproofing that could prove useful in our function. Run this command at the prompt:

>> *terminalVelocityDebug('mercury')*

The output here is inf, which you hopefully remember is the MATLAB constant for infinity. Mercury has essentially no atmosphere at all, so its atmospheric density is 0. Because *atmosphericDensity* is in the denominator of the terminal velocity calculation, we're dividing by zero. MATLAB returns inf by definition when dividing by zero, but many programming languages result in an error. A try/catch block lets you fail gracefully when that happens rather than crashing with a system-generated error.

The **try/catch** block uses this format:

```
    try
        % Code that could generate an error
    catch
        % Code to handle the error, if one occurs
    end
```

If no error is caused by the code in the **try** block, the **catch** block is skipped and program flow continues normally. If the code in the **try** block does cause an error, the code in the **catch** block is executed.

The error-handling code in the **catch** block can be anything you need. It can be a function-defined error message to let the caller know what happened (including variable values at the time), and a function termination. It could be an **if-ifelse-then** block that responds differently depending on the error. It could assign a new value to a variable that caused the error and continue the function. It can be anything you want; it's a flexible way for you to proactively confront an otherwise catastrophic event in your code.

This is a somewhat invented example, but let's pretend that MATLAB would crash upon attempting to divide by zero.

> ***On your own, in the Editor window, enclose the terminal velocity equation in a*** try/catch ***block as shown:***

```
try

    Vt = sqrt((2 * mass * gravity) / (atmosphericDensity *
      C * area));

catch

    disp("Error - function terminated");

    return;

end
```

Add the **try/catch** technique to your debugging toolbox and consult the MATLAB help if you want to learn the more sophisticated technique of capturing and responding to specific error messages.

As we've said, debugging is an art, and an arcane one at that. There is no single right answer to finding a problem in code; entire classes have been taught and books written about debugging. We can't cover it all, but here are a few general techniques used by a lot of programmers.

Comment Problematic Lines

Temporarily making a line into a comment (commonly called *commenting out*) is sometimes useful when you're debugging code. There are two comment formats beyond the standard that are also helpful in troubleshooting:

- Use two percent signs (%%) to specify a section title. Titled sections can be run independently.

- Use the %{ and %} strings to enclose and comment out an entire block of code without having to comment out each line individually. This is an easier method if you want to temporarily skip a large section. These strings must each get their own line, as shown here.

```
x = x + 77;
    %{
        z = cos(88.4);
        A = magic(3);
    %}
    plot(x);
```

To comment out a large block of code, you can also highlight the lines and then hit the [Ctrl] + [R] key combination to comment them all out at once. [Ctrl] + [T] uncomments them.

Tip: *To see the full list of keyboard shortcuts for various actions and ribbon commands, open the* **Preferences,** *found in the* **Environment** *toolbar in the* **Home** *tab. You'll find the list of shortcuts under the* **Keyboard** *category. Professional developers make heavy use of keyboard shortcuts for their efficiency.*

Remove Semi-Colons

As much as MATLAB urges you to suppress the output of individual lines in a script or function, in some cases it can be helpful to see that output as it runs. Take any number of those semi-colons off temporarily if it's of use to you.

Hover to See a Value

In a large program, you may have more variables than you can see at once in the Workspace window. Hover your cursor over a variable name anywhere that it appears in the code, and a pop-up shows you that variable's current type and value. Note that it might not have been that variable's value at that specific line in the code, but instead is its value at the line where the function is paused.

Experiment with this on your own in the paused *terminalVelocityDebug* function.

Enter Your Own Values

Remember that you're in the scope of the function's workspace (K>>) in the Command Window, but otherwise it's the same old Command Window it's always been.

If a calculation in the function is giving you a wrong value, you can set a breakpoint just after that calculation, assign that variable a correct value through the Command Window, and proceed to see if the rest of the function works. The function variables in this workspace can be manipulated at the Command Window prompt just like any other variable and you can use this feature to experiment with different values.

Pause Running Code

When a function runs normally without any breakpoints, the Run button becomes a Pause button during the function's execution.

Figure 7.13

Often, a function runs so fast that only The Flash from comic books would be quick enough to pause it – you can't even see the button change from Run to Pause and back again – but some take longer to run or are even designed to do so. The Pause button acts as an on-the-fly breakpoint. Once paused, the function halts at its current *point of execution* and the Debug toolbar replaces the Run toolbar.

To experience breaking out of an infinite loop make this temporary change to your code:

On your own, add this infinite loop at the top of the terminalVelocityDebug function

```
while 1
    x = 1;
end
```

Now run the function

The loop ensures that the function just sits there.

Click: **Pause button**

This opens the debugger with the green arrow pointing at the current line. If you were debugging this code, you would see that this is where it was hanging up, and why. This is also a convenient way to get out of an infinite loop.

Click: **Quit Debugging to exit the function.**

Testing Your Code

You must test your code to know that you've produced a working function or program. This doesn't mean just throwing a couple values at it and calling it a success because the right values were returned. Put real thought into devising challenging cases, inventing bad input, and pushing the boundaries of the values the code confronts. You don't simply want code that works perfectly with well-mannered data, but code that can take care of itself in the binary equivalent of a seedy waterfront bar in a pirate town. Do everything you can to make your code fail, fix it when it does, and then make it fail again.

Remember that for a program to work correctly, each piece of it must work correctly. It's harder to track down problems in entire programs than it is in single functions, so do everything you can to make your functions infallible. No developer or group of developers can foresee every possible failure condition in the often complex and lengthy code used in professional software, but with thorough testing, your code has the best chance of success. Be assured the world will throw something unexpected at your code.

You have finished Tutorial 7.

Key Terms

debug

debugger

error handling

semantic error

syntax

Key Commands

%%	[Ctrl] + [T]	dbstep in	return
%{　%}	dbquit	dbstep out	try/catch
[Ctrl] + [R]	dbstep	lower	upper

Exercises

 Exercise 7.1

Write a function with a single input parameter that is a 3 x 4 matrix that contains either the twelve signs of the Chinese zodiac, the twelve signs of the western zodiac, or a mix of the two, as strings (double-quotes). Write code in the function that verifies that the matrix doesn't contain any strings outside of the expected set and that it contains no duplicate elements.

 Exercise 7.2

Open the data file, *terminalDebugging.m.* Use the debugging skills you have learned to debug and run this code.

 Exercise 7.3

Open the file *secretMessage.m* included in the download. Be sure that the secretMessage.dat file is also in the same directory. Run the function by entering the following command in the Command Window:

>> *secretMessage('6')*

The function should fail to run. Using the techniques learned in this chapter, track down the function's many flaws until running the function reveals the encoded text.

Two hints: the text begins "Four score…" and even the function call given above is suspect.

IMPORTING AND EXPORTING DATA

8

Introduction

In post-education working life, you don't spend much time keying in X = 5. Instead, there's real world data to explore, analyze, and mine: seismic data, stress data, polling data, sub-atomic particle collision data, sales data, astronomical data, population data…the list is endless. The scientists, engineers, sensors, and devices that gather data make it available in a wide variety of formats: text files, databases, spreadsheets, binary streams, XML, etc. That data can be brought into MATLAB to be organized, visualized, manipulated, and analyzed. The results can then be given back to the world in a format others can use.

MATLAB provides a rich pool of functions to import and export data, plus an interactive import tool that simplifies the process by allowing you to select data visually. You can import text, spreadsheets, images, audio, video, and dedicated scientific data in file formats such as *.txt*, *.jpg*, *.mp3*, *.avi*, and *.xlsx*. We will be covering the basics of importing and exporting delimited text files and Excel spreadsheets, but there is so much more.

Navigating the File System

Data is stored in *files*. Files are stored in *folders* (also known as *directories*). Folders are stored in parent folders. At the tip of the storage pyramid is the *root* folder, which in Windows is a particular hard drive, such as *C*. The folder hierarchy from the root to the file is called its *path*, for example *C:\Users\Public\Documents\MATLAB\matlab.mat*.

Files and folders can be created, of course – we've created lots of files already, and they can also be moved, deleted, copied, or renamed.

Where Am I?

The folder whose contents MATLAB shows in the Current Folder window is also called the *working directory*, and its path is shown in the bar above the Current Folder window. See Figure 8.1.

You can also see the current directory path by entering the pwd command in the Command Window.

Try that now, assigning that path to a variable.

>> *curr_dir = pwd*

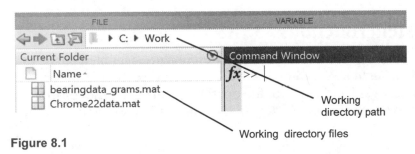

Figure 8.1

Objectives

When you have completed this tutorial, you will be able to

1. Understand the Windows file system hierarchy with files and folders

2. Navigate the file system through the command prompt

3. Copy and move files and folders from the command prompt

4. Understand the difference between absolute and relative paths and construct a relative path to a file

5. Use wildcards in a path to refer to multiple files or folders

6. Save variables to and load variables from a .mat file

7. Use functions to import data from an Excel spreadsheet and a delimited text file

8. Use the MATLAB import tool to import data from an external file

9. Use functions to export data to an Excel spreadsheet, a comma-separated value file, and a text file

So far, we've stored all our script and function files in *c:\Work* which we set as MATLAB's default working directory (the one it starts in). See the chapter on Preparing MATLAB for the Tutorials. To import and export data, you need to understand how to find the import file and where the exported file is being placed. This may be rudimentary for many of you, but it is important to understand, so we include it here.

When you call a function or script from the Command Window, MATLAB first looks for a matching file in the current directory. If no match is found, MATLAB works its way through several other folders known to it. To see the full list of folders searched, type this command:

>> *path*

To access a file that is not stored in one of the search path folders, you must either navigate to its location or refer to it in commands by its full path name.

A path name can be given in two forms, *absolute* and *relative*. An absolute path gives the full path from the drive root:

'C:\Users\Public\Documents\Matlab'

The drive letter can be omitted if you are referring to a path on the same drive as the reference. Begin the path with a slash to indicate the root.

'\Users\Public\Documents\Matlab'

A *relative* path makes use of the special **..** directory indicator which means parent of the current folder and is present in all folders except the root (which by definition has no parent). For instance, assume that you're in the working directory *C:\Users\Public\Documents\Matlab*. Enter the following command to list the files in the Music folder, which is a Windows default and a sibling of the *Documents* folder.

ls ..\..\Music

Got it? The first '..' takes you up one level to the *Documents* folder. The second '..' takes you up another level to the *Public* folder. The Music folder is stored in the *Public* folder, so its contents are listed. On your own, navigate around a bit from the Command Window prompt, using both folder names and the '..' indicator. Notice how the content in the Current Folder window changes as you do so.

Relative paths are useful in programming, especially for referencing your own content. A user can install your application to a non-standard location, but because you determine your app's internal structure, you can use relative paths to access it regardless of its parent folder.

Paths are *not* case-sensitive.

File System Functions

MATLAB provides all the basic file system functions familiar to anyone who has ever used a Windows or UNIX command prompt. You can choose to navigate and perform file operations through functions, or with the mouse by clicking folders in the Current Folder window or by

Tip: *MATLAB accepts either the backslash (\) or the forward slash (/) in paths. You can even mix them in the same path (but don't).*

using the Up One Level or Browse For Folder buttons. However, in code, you have to use the functions.

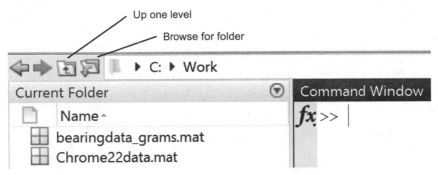

Figure 8.2

Most of the functions described in this section have two forms: a command form and a function form. The command form is used in the Command Window and takes this general form:

> someFunc *parameters*

The function form is used when the return value is assigned to a variable. It requires that the parameter(s) be enclosed in parentheses. This is the general function form of the command above:

> retVal = someFunc(*parameters*)

Wildcards

Many of the commands we'll discuss accept an *asterisk (*)* as part of the path. This is called a *wildcard* and is used to represent "all". Some examples should make this concept clearer.

copyfile ('*', 'C:\Users\Public\Documents\Matlab2') says to copy everything in the current folder (both files and folders) to a folder called *Matlab2*.

copyfile ('*.txt', '..\Matlab2') tells MATLAB to copy all files with the extension *.txt* in the current folder to a folder called Matlab2.

cd

Short for *change directory*, the **cd** *path* command moves from the current directory to the folder specified in the *path* parameter, which becomes the new current directory. The path can be absolute or relative, but can't contain *wildcards*. If the path contains a space, it must be enclosed in single quotes. The path cannot include a file name.

Type the following command and note the result in the path bar above the Current Folder window.

> >> **cd \users\public**

Next, type this command to move from there to the *Users* folder:

> >> **cd ..**

The **cd** .. command is the equivalent of the Up One Level button.

Tip: *Along with the '..' indicator, you'll see the '.' indicator, which means "the current directory." Because this doesn't take you anywhere except where you already are, it isn't used as much.*

Tip: *For the full list of recognized import formats, see the Supported File Formats for Import and Export topic in the MATLAB Help system.*

If you use a string variable to specify a path to the *cd* command, the variable must be in parentheses as shown here:

cd (*pathVar*)

ls/dir

Use ls to list the contents (files and folders) in a specified folder. This is roughly equivalent to the Windows *dir* command or the UNIX *ls* command.

- ls lists the contents of the current folder.

- ls *folderName* lists the contents of the specified folder if it is present in the current folder. If the folder name contains a space, it must be enclosed in single quotes.

- ls **path** lists the contents of the specified folder. If the folder name contains a space, it must be enclosed in single quotes.

If you assign the result of an ls command to a variable, the folder name or path must be enclosed in parentheses as shown here:

contents = ls ('\Users\Public\Matlab\My Code')

The variable receives the list as an array of characters. Each row is a single file or folder name and all rows are the length of the longest file name. Row with shorter names are padded with spaces at the end to make up the length. Type the following command to see this:

files = ls

The MATLAB dir command is equivalent to the ls command until you assign the result to a variable. The dir command returns an array of structures, one structure for each file. Each structure contains further information about the file. The next table describes the information fields.

Field Name	Data Type	Use
name	character array	The name of the item
folder	character array	The absolute path of the folder that contains this item
date	character array	The date and time that the item was last modified
bytes	double	The size of the item in bytes
isdir	logical	1 if the item is a folder; 0 if the item is a file
datenum	double	The date and time that the item was last modified, as a number

mkdir

Use mkdir to create a new directory. There are several parameter options.

- mkdir *folderName*: Creates the folder in the current directory. If

Tip: *Unlike the Windows cd command, you must leave a space before the '..' specifier.*

Tip: *To be clear,* ls *is short for "list" and is the lower case of LS. On the page, it can look like the capitalized word "is."*

there is a space in the folder name, it must be enclosed in single quotes.

- mkdir *path\folderName*: Creates the new folder in the directory specified by the path. The path can be absolute or relative.
- mkdir *path folderName*: As above, but supplying the path as a separate parameter from the folder name. This form allows more flexibility when you're using the command in code.

The mkdir function has three return parameters that tell you how it went: [*status, msg, msgID*]

- *status*: 1 if the directory was created successfully; 0 otherwise.
- *msg*: if the status value is 0, this is the text of the error that occurred to prevent creation of the folder.
- *msgID*: the ID of the error message. This can be used in a try/catch block.

Type the following command assigned to a variable named *status*:

>> **status = mkdir ('..\NewFolder')**

The new folder is created and the mkdir function returns 1 to indicate success. Now issue this command:

>> **[status, msg, error] = mkdir ('..\NewFolder?')**

That question mark makes for an illegal file name, so the operation fails. Because you asked for the returned information, the text of the error message and error code are returned with the status of 0.

rmdir

The rmdir *folderName* command deletes an empty folder. If the folder is not empty, the operation fails. The rmdir function returns the same three values as mkdir which are status, error message, & error message ID.

Issue the following command to delete the directory that you created above.

>> **[status, msg, error] = rmdir('..\NewFolder')**

delete

The delete *fileName* command removes the specified file. As with other commands, the *fileName* parameter can be just the file name, in which case it must be in the current directory, or specified with path information if it is in a different location. This command cannot be used with folders. Whether the deleted file is permanently erased from the file system or is moved to a trash or recycle bin folder from which it can be retrieved depends on the *Deleting files* setting found in the **Preferences** under *General*. You can either change the setting in **Preferences** or programmatically though the recycle command as shown here:

```
recycle('on')
```

movefile

The **movefile** *source destination* function moves a file or folder from one folder location to another, optionally renaming it in the process. Moving a folder also moves its contents.

The original *source* file or folder name, either by itself if the file or folder is in the current working directory, or with additional path information. You can include *wildcards* in the file name (such as *.mat) to move multiple files in one call.

The *destination* final file or folder name. The destination cannot include wildcards. If you omit the destination, the current folder is used.

- If *source* refers to a file and destination refers to a folder, the file is moved into that folder.

- If *source* refers to a file and destination refers to both a path and file name, the file is moved into the new folder and is given the new file name.

- If *source* refers to a folder and destination refers to an existing folder, the original folder is moved into the destination folder

- If *source* refers to a folder and destination refers to a new folder, the original folder is moved to the new location and given the new folder name.

The **movefile** function is also used to rename a file. If the source and destination folders are the same, the source item is just given the new name.

The **movefile** function returns the familiar status and error values: [status, msg, msgID]

copyfile

With the exception of renaming, the **copyfile** function works just like the **movefile** function, but leaves the original file or folder in place.

Let's put these commands together to give you some experience. Enter these commands (without the comments) and observe the results.

```
>> home = 'c:\users\public\documents\matlab'
>> cd(home)          % Start from the default working directory
>> pwd               % Assure yourself you're in the right place
>> ls                % List the files in this directory
>> edit testfile1    % Make some files to play with
>> edit testfile2
>> edit testfile3.txt
>> mkdir ..\matlab2  % Create a new sibling directory
>> cd ..
>> pwd               % Where you are now
>> ls                % Notice that matlab2 exists here
```

```
>> cd matlab2          % Move into that directory
>> ls                  % Nothing here yet

>> copyfile('..\matlab\testfile*.m') % Copy files to this folder
>> ls                  % And there they are
>> movefile('testfile1.m', 'testfile4.m')  % Rename a file
>> dir                 % Verify the name change
>> cd(home)
>> pwd                 % And you're back
>> %rmdir('..\matlab2')     % Can't do it – the directory isn't empty
>> delete('..\matlab2\*.*')         % Delete the copied files
>> rmdir('..\matlab2')              % Works now
>> delete('testfile*')              % Clean up
```

Other Functions

Here are a few other commands that you'll find useful in regard to navigation, paths, and files.

which

Searches the MATLAB search path for a specified file or folder. If the item is found, its path is returned. For instance, *pathdef.m* is the file that contains the directories list for the MATLAB search path. Type the following command to see where it's stored:

```
>> which pathdef.m
```

uigetdir/uigetfile

These commands present the user with a standard file dialog window from which they can select a file or folder to return its path. This is useful for retrieving an item's path to include in a variable or code, or to make the code interactive.

Type this command in the Command Window to see how it works. Choose any folder to see the result.

```
>> uigetdir
```

type

The type function shows a file's contents. It can't make sense out of every file type and in those cases returns what looks like gibberish, but it's fine with text files and some MATLAB formats such as *.m*. Unfortunately, it doesn't work with .mat files.

whos

Lists the variables in the Workspace or in a file, together with their types

Tip: *Use the* which *command to find source files for MATLAB's built-in functions, which are .m files just like any other function. Looking at their source code can help you write your own code.*

and values. The command alone lists the variables in the Workspace.

whos

By adding the **-file** flag and a file name, the **whos** function lists the variables held in that file.

whos -file matlab.mat

The **whos** command can also list all the global variables by using the global keyword.

whos global

Working with .mat Files

In the first pages of this book, you exported the contents of the Workspace to a *.mat* file, a MATLAB-specific file format that holds variables and their values. Later, you brought those variables back into MATLAB by importing that same .mat file. We'll review briefly here, since this is the simplest case of importing data.

When you exit MATLAB, the variables in the Workspace are erased from the device's memory. The Workspace is empty when you restart the program. To save those variables, you can do so in a *.mat* file. Type the following lines into the Command Window:

>> *clear   % Start with an empty Workspace*

>> *a = 1*

>> *b = 2*

>> *save*

The **save** command creates the file *matlab.mat* and stores it in the current directory.

Now type the following:

>> *clear*

>> *load*

The **clear** command removed all the variables from the Workspace and the **load** command restored them from the saved file. If you enter the **load** command without specifying a .mat file name, it looks in the current directory for a file named *matlab.mat*. Type the following commands:

>> *delete matlab.mat*

>> *load*

The **load** command gave you a "*no such file*" error because there is no *matlab.mat* file present in the current directory, and that's as far as it's going to look.

Type the following to put some variables back into the Workspace:

>> *a = 1*

>> *b = 2*

>> *save*

All is as before. Now type these commands:

> *>> c = 3*

> *>> save*

The earlier *matlab.mat* file is overwritten without a warning. That could be catastrophic if the earlier *matlab.mat* file was from a previous MATLAB session. Instead of simply typing the save command, you can give the file any valid name you'd like to store a snapshot of your session.

> *>> d = 4*

> *>> save vars.mat*

You can save the Workspace of a running function too (recall that functions use their own Workspace). To do this, set a breakpoint at a line in the function where the variables you want to save have been defined. While the function is paused, enter the **save** command (with your own file name, if desired) at the K>> prompt. You can load those variables back into the base workspace if you'd like; a *.mat* file doesn't have to be loaded to the environment it was saved in.

If you already have some variables in your Workspace and you load a *.mat* file, the variables in the *.mat* file are added to the existing Workspace variables, which remain unchanged. Unchanged, that is, unless an existing variable has the same name as a variable in the *.mat* file. In that case, the existing variable is overwritten by the .mat version with no warning.

Type the following commands:

> *>> d = 1*

> *>> load vars.mat*

The variable *d* stored in the *vars.mat* file overwrites the variable *d* in the Workspace, changing its value from 1 to 4. The obvious follow-up question is "how do I know what variables are in the .mat file?" Because a *.mat* file is not human-readable (prove it to yourself if you want by typing **type matlab.mat** at the prompt), you have to use the **whos** command. Type the following command to reveal the variables held in our example *.mat* file:

> *>> whos -file vars.mat*

When you know the variable names in a *.mat* file, you have the option of loading only select variables instead of all of them. List the variables you want at the end of the **load** command, separated by spaces. Enter the following commands:

> *>> clear*

> *>> load vars.mat a c*

The Workspace should now contain only variables *a* and *c*, loaded from the *vars.mat* file.

This same technique works with the **save** command. Enter the next commands to demonstrate this:

Tip: *To load a .mat file, you can also just double-click that file in the Current Folder window.*

```
>> b = 2
>> d = 4
>> save vars2.mat c d
>> clear
>> load vars2
```

Importing External Data

External data can come in many forms. It might be stored in a spreadsheet or a database or a text file or just as a bunch of 1s and 0s. It can also be stored anywhere on your device or a network (local or cloud) that your device is part of. It might not even be stored, but instead be streaming from the device. Given all these possibilities, we'll look at some of the facilities that MATLAB provides for bringing that data in for you to use.

Excel Spreadsheets

Perhaps unexpectedly, one of the easiest sources of external data for MATLAB to import is a Microsoft Excel spreadsheet. What is a spreadsheet? Each sheet in a workbook is a grid of rows and columns; a matrix, in other words. If the spreadsheet contains multiple sheets, it can become a three-dimensional array. It's a natural fit for MATLAB.

Because of the ubiquity of Excel in industry and science, plus the ability of a wide variety of programs to read Excel data, MATLAB provides special functions for the import of data from and export of data to Excel.

xlsread

The xlsread function imports data from an Excel spreadsheet into MATLAB.

Here is the basic form of the function call:

```
[num_cells, txt_cells, all_cells] = xlsread ('fileName');
```

Notice the semi-colon. With even a small spreadsheet, a lot of information is echoed to the Command Window, so it's best to always suppress it.

The return variables, which we've named *num_cells*, *txt_cells*, and *all_cells* (but which you can name anything you want), are three views of the same data taken from the spreadsheet.

The first return variable (we called *num_cells*) receives the smallest matrix that contains all the numerical data in the spreadsheet, plus any other cells included in that area. This makes more sense visually as shown next.

Tip: *Technically everything stored on a digital device is 1s and 0s, but in this case we're defining a binary file as a non-text file or stream without the inherent structure provided by a spreadsheet.*

Imagine a spreadsheet that contains the cells shown to the right

120	abc	def
ghi	180	jkl
360		

MATLAB determines the smallest rectangle that encloses all the numeric data and returns that as a matrix.

120	abc	def
ghi	180	jkl
360		

Numbers are imported as type double. Cells within the rectangle that do not contain numeric data (including blank cells) are given the MATLAB constant value NaN, for "not a number."

120	NaN
NaN	180
360	NaN

The second return value (*txt_cells*) does the same thing for text, again finding the smallest rectangle that can lasso all the text cells.

120	abc	def
ghi	180	jkl
360		

The result is a cell array. Cells that do not contain text (including blank cells) are given an empty string ('').

''	abc	def
ghi	''	jkl

The last return value(*all_cells*) retrieves all data on the spreadsheet, again using the smallest rectangle that can surround it.

120	abc	def
ghi	180	jkl
360		

The data is returned as a cell array. Numbers are stored as doubles, text is stored as character arrays, and empty cells are given the value NaN.

120	'abc'	'def'
'ghi'	180	'jkl'
360	NaN	NaN

Tip: *Want to know if a name you want to use for a variable or function is already used by a built-in function? Use the exist command, such as* `exist sine`. *See the MATLAB help for more details, but basically, if it returns 0 you're safe.*

As with any function that returns multiple values, you don't have to receive all of the returned information. You can omit return variables either by leaving variables off the end or replacing them with a tilde (~). This command retrieves only numbers because that's the first return variable:

```
num_cells = xlsread ('path\fileName');
```

This command retrieves only text data:

```
[~, txt_cells] = xlsread ('path\fileName');
```

Using further **xlsread** function parameters, you can specify which sheet of an Excel workbook the data will be read from and/or limit the range of cells to retrieve.

We've provided the sample Excel spreadsheet *IO Example.xlsx* with the data files for you to use.

Issue the following commands, replacing the example path with the path of your *IO Example.xlsx* data file if you've stored it elsewhere:

>> *path = 'C:\datafiles-matlab\IO Example.xlsx';*

>> *sheet = 2;*

>> *range = 'A1:E10';*

>> *[num_cells, txt_cells, all_cells] = xlsread(path, sheet, range);*

The range value is a character array that uses the Excel cell naming convention to describe the rectangle of cells to import. The cell to the left of the colon is the upper left corner and the cell to the right of the colon is the lower right corner. Remember that the colon here is not the MATLAB colon operator, but just a normal colon.

Sheet numbers are 1-based, so 1 is the first sheet, 2 is the second, and so on. Alternately, you can use a sheet's name instead of its number.

Not sure which cells you'll want ahead of time? Issue this command:

>> *[num_cells, txt_cells, all_cells] = xlsread (path, -1);*

In response, MATLAB displays a message telling you to switch to Excel (this obviously doesn't work if Excel isn't installed on that machine) and manually select the range of cells. This is limited to a continuous block of data; you can't pick a collection of individual cells.

Text Files

A text file is a free-form, human-readable file that contains only character data. A Microsoft Word document is not a text file because it contains quite a bit of information other than the words on the page – formatting data, etc. On the other hand, a Windows Notepad document (*.txt*) is a text file – simple and unadorned.

A text file can contain numbers, of course, but those numbers are stored as text. The text can also be laid out in a meaningful way, such as in columns or with a specific character as a delimiter between items. It generally requires more knowledge about the source file to import a text file than it does to import an Excel file.

Opening a File

To use certain functions with a text file, you must explicitly first open the file with the **fopen** function. The command takes this form:

fileID = fopen ('filename', 'permission')

The resulting file identifier returned by **fopen** is a positive integer value 3 or greater that you'll use in other file operations, such as closing the file. A value of 0 is reserved for standard input, 1 for standard output to the screen, and 2 for standard error.

If the value returned is -1, the file could not be opened. Always check this value to verify that the file opened correctly before proceeding in your code, because your program will crash if it attempts to access a closed file. It is sufficient to check whether the fileID is greater than 0.

The *permission* parameter is a string (single- or double-quotes) that determines what you can do with the file once it's opened. See the table below for the values that can be assigned for permissions.

Permission	Meaning
rt	Open for reading only, contents cannot be changed. This is the default value if you omit the *permissions* parameter.
wt	Open for writing, deleting all existing content
at	Open for writing, appending any new content to the end of existing content
rt+	Open for both reading and writing. Does not create a new file if the specified file does not exist.
wt+	Open for both reading and writing, deleting all existing content. Creates a new file if the specified file does not exist.
at+	Open for both reading and writing, appending any new content to the end of existing content. Creates a new file if the specified file does not exist.

Enter the next command, which will create a new text file in the current working directory. You must provide the file extension (here, '*.txt*') as part of the file name.

>> ***fileID = fopen('MyNewFile.txt', 'wt+')***

You see fileID = 4 echoed in the Command Window.

Closing a File

Once you've opened a file, you need to close it again when you're done. What you just read there is vital information – read it again. Failure to close a file can lead to its being locked to other programs or users, memory leaks, or even system crashes. When you're importing data, the best practice is to open the file, read the data into MATLAB, and close the file immediately.

Closing a file is simple, using the **fclose** function, for example, fclose (*fileID*). The optional return value has a value of 0 on success, -1 otherwise.

Not sure you've closed everything? Call the **fopen** function with the **all** keyword. Enter the following command in the Command Window:

>> ***ids = fopen ('all')***

This returns a row vector of all open fileIDs, or an empty matrix if no files are open. In code, you could then loop through the vector, calling fclose on each value. We've only opened one file so far, so you should see it stored as *fileID* from the previous command entries. Next close that file (if you didn't save the file identifier as *fileID*, substitute your file's ID).

Tip: *A memory leak occurs when memory resources aren't released back to the system once an application is done with them. This can pile up until you're out of memory or storage space and a crash occurs.*

```
>> fclose (fileID)
```

The 'all' keyword can also be used with the fclose function to close all open files.

```
>> fclose ('all')
```

Reading a File

Reading a text file can be a problem because it doesn't have the enforced grid of a spreadsheet. Where does one item end and the next begin? Is each line a different length? Are there line breaks at all? Is the text in a row- or column-based form that you should preserve or that helps you interpret the data? MATLAB provides several functions to do so based on the file's organization.

csvread

The csvread function is used with comma-separated values, a common text file format. It does not require the file be opened. Comma-separated value (*csv*) files often have an extension of *.csv*, but it's not a requirement. The csvread function can only be used to read numerical data.

Because of its layout, a *csv* file can be broadly thought of as a grid. Each row in the *csv* file corresponds to a row in matrix, and each item between commas corresponds to one column entry. The data file *CSVtest.csv* is a comma delimited file that contains the following lines. You can open *csv* files with Notepad or a similar app to view them.

```
1,3,5

2,4,

7,8,9
```

Enter the following statement in the Command Window, replacing the path if necessary to where you have stored your data files.

```
>> csvread ('C:\datafiles-matlab\CSVtest.csv')
```

The result in the command window is the following matrix:

```
ans =

   1   3   5
   2   4   0
   7   8   9
```

You can specify individual elements or ranges in the csv file, in case you only want to read part of the file, but unlike every other thing we've done up until now, the first row and column in a *csv* file have an index of 0, not 1. The numeral 2 in the example above is in row 1, column 0.

The csvread function outputs a matrix, in keeping with the spreadsheet analogy. If the rows in the *csv* file are of different lengths, or if values are skipped (such as '1,2,,3,4'), 0s are added as necessary to fill the matrix.

We've also supplied the *csvtext.csv* file for your experimentation. Enter the following command, adding path information to the file name if you haven't stored the file in the current directory.

```
>> csvContent = csvread('csvtest.csv');
```

That reads the entire file. Now import the data again, skipping the first line (a good technique to know, since the first line in a *csv* file is sometimes text that supplies column names).

```
>> csvContent2 = csvread('csvtest.csv', 1, 0);
```

The first number after the file name is the row from which to start reading, and the second number is the column (normal MATLAB order).

Specifying a range of values has an odd format. Enter this command:

```
>> csvContent3 = csvread('csvtest.csv', 1, 1, [1,1,3,4]);
```

The numbers in brackets represent the upper-left and lower-right corners of the rectangle of values to read, in this case (1,1) and (3,4). The numbers before the bracket are again the row and column from which to start reading, but because that's also given with the upper-left corner values in the brackets, the end result is that the row and column numbers are always the same as the first two values in the brackets.

dlmread

The dlmread function is just a general form of the csvread function, allowing characters other than a comma as a delimiter. Again, it does not require the file to be opened first, the data read from the file can only be numbers, and 0s are added as necessary to fill the matrix.

If you call dlmread without specifying a delimiting character, the function attempts to deduce it from the file's contents. Try this command with the *dlmtext.dat* file that we've provided, which uses the semi-colon as a delimiter:

```
>> dlmContent = dlmread('dlmtest.dat');
```

To avoid misinterpretation, you can explicitly specify a delimiter. This command should give results identical to the command above:

```
>> dlmContent2 = dlmread('dlmtest.dat', ';');
```

If you used dlmread with a *csv* file and a delimiter of ',' you would do just what the csvread function does behind the scenes: it calls dlmread to do the real work.

Just like csvread, you can specify that dlmread begins reading at a certain point in the text file. However, there is a difference in specifying a range of values. The row and column values that precede the brackets are no longer required, and the delimiter must be specified. Enter this command

```
>> dlmContent3 = dlmread('dlmtest.dat', ';', [1,1,3,4]);
```

Using the Import Data Tool

Because importing data can sometimes be complicated, MATLAB provides an interactive tool that not only allows you to visually select the data you want to import, it writes code for you to use next time! This is not only a great importation tool, it's also a great learning tool.

Can't figure out the code to include in a function that needs to go out and retrieve data? Do it manually with the Import Data tool and copy the code that it generates to your function.

Remember the Import Data tool on the ribbon's Home tab?

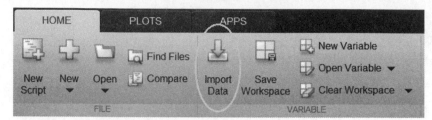

You can also right-click a data file in the Current Folder window and select Import Data… from the context menu.

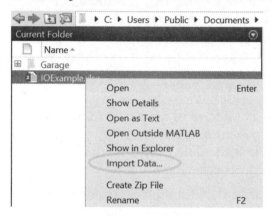

If you click the Import Data item on the toolbar, you are presented with a standard file dialog window from which to choose your data file. Locate the *IOExample.xlsx* file through the file dialog window and select it. There may be noticeable pause while MATLAB gathers the information, and then you're presented with the Import page.

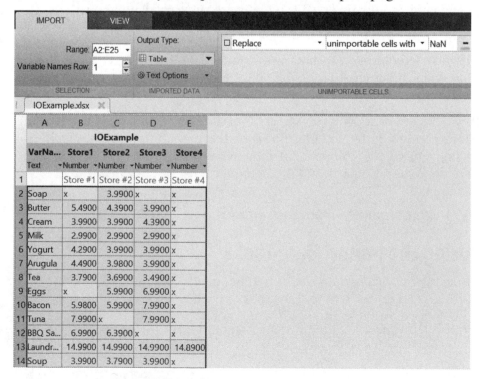

Options

The toolbars in the ribbon let you customize the import operation.

Range lets you enter a range of cells or choose from a selection of previous ranges (if applicable). When you choose cells in the data itself, the selection is reflected here.

Variable Names Row is available only when the output type is a table or column vectors. This option specifies that the text in the given row should act as column headers. In the grid where you can select the data to import, you can also specify the name of the imported matrix or vectors. In the case of column vectors, you can specify the name of each.

The **Output Type** drop-down menu shows the destination type for the imported data. The default is Table, a data type we haven't talked about yet. The other options are:

- **Column vectors**: Each column is imported as a separate variable, each containing a single column vector of a single type

- **Numeric Matrix**: Imports only numbers from the source data

- **String Array**: Imports all the selected cells as strings

- **Cell Array**: Imports all the selected cells into a cell array. Numbers remain numbers and text remains text.

Text Options allows you to specify how strings should be stored: as the string type (double quotes) or as character arrays (single quotes).

Unimportable Cells lets you set rules that handle data that can't be imported as a certain type. MATLAB color-codes the cells in the data to show you which can't be imported under the current import type: blue for importable, yellow for unimportable. The default rule for an unimportable cell replaces that cell content with the NaN value. Other options are to exclude any column or row that has at least one unimportable cell.

This option is not available for the String Array or Cell Array, because those types don't have any unimportable cells.

Import Data imports the data using the options that you've chosen. The new variable(s) appear in the Workspace.

Generate Script and **Generate Function** are where things get interesting. Instead of importing the data, they generate the code to import the data, which you can save and run to perform the actual importing now and in the future. The generated code is profusely commented, so it gives you everything you need to know about how to use it, what it's doing, and why. Apart from being a handy feature, this is also a splendid learning tool. Being able to compare code that you've written to accomplish a task with auto-generated code can open your eyes to possibilities that hadn't occurred to you (including the fact that your code might be better).

Exporting Data

Once you've analyzed data in MATLAB, you can export it in various formats for use by other people.

Exporting to Excel Spreadsheets

MATLAB can export data directly as an Excel spreadsheet on a Windows system with Excel installed. The export operation is straightforward and mirrors the import operation.

The xlswrite function is called as shown here:

[status, message] = xlswrite (*filename,arrayname*)

The *filename* parameter is the destination file. To write to the current directory, just give the file's name. To write to another directory, include a relative or absolute path as part of the file name. If the file does not exist, it will be created.

The *arrayname* parameter is the MATLAB array that you're exporting. Cells with a value of NaN or empty strings are written as blank cells in the spreadsheet.

The return value *status* tells you whether the operation was a success. A value of 1 means everything went great. A value of 0 means that there was a problem.

The return value *message* gives you the error message text if an error was generated by the operation (*status* = 0).

As with xlsread, you can also specify the sheet and/or cells to write to in the destination file. When writing to an existing file, be aware that if there is already data in the cell rectangle that you're writing to, that data will be overwritten and lost with no warning. You also can't write to a currently open spreadsheet.

Exporting Text

In some cases, a spreadsheet won't fit your needs or those of the consumer of your data. In those situations, a text file – as something simpler and even more universal – is the better option. We'll talk about two methods of output: csv files and formatted text.

Exporting to CSV Files

MATLAB provides the *csvwrite* function to export data in the comma-separated value format.

```
csvwrite ('filename', matrix)
```

Simply provide a file name, as always, just the file name to create the file in the current folder. You can provide a relative or absolute path before the file name to create it in the location you specify. Note that if a file with that name and location already exists, **csvwrite** will overwrite it without so much as a "*by your leave!*" The **csvwrite** function has no return value.

Exporting Formatted Text

You've already had a little exposure earlier in the book to the **fprintf** function, which sent formatted text to the Command Window. The only difference in exporting that same text to a file is that you provide **fprintf** with the file's ID that you got from the **fopen** function.

```
fprintf (fileID, 'format specifier', var1, var2, …)
```

The key to **fprintf** is the *format specifier* parameter. This is a string (either single or double quotes) that contains both literal text and placeholders for variable values. There are different placeholders for different data types, which affect how the data looks when it is written to the screen or to a file. Try this simple example in the Command Window:

>> *fprintf ('My name is %s.\n', name);*

The **%s** in the format string is a placeholder for a string variable, the value of which is provided by the *name* variable (it could also be a literal string). The rest of the format string is output as given, including spaces. The \n character provides a new line (carriage return) at the end of the line; otherwise, the next output would begin immediately following the period. Almost all of your format strings will end with the \n character (and yes, it is regarded as a single character).

Here are the most common type specifiers used in a format string:

Type Specifier	Meaning
%s	string
%d or %i	signed integer
%u	unsigned integer
%f	floating-point number

There are many more, which you can find in the MATLAB Help topic on **fprintf**.

The floating-point specifier can specify the number of digits to output. For instance, **%.2f** prints two digits to the left of the decimal point.

The \n character, which means *new line*, is what's known as an escape (or escaped) character. This and its kin allow you to use characters in a string that would otherwise be interpreted non-literally. For instance,

Tip: *On a Windows system, if Excel is not installed, xlswrite can only export numeric data, and only as CSV text file. That's also the case on a Mac, even with Excel installed. CSV files can be read by Excel, so Excel can still access the data that way.*

Tip: *If your data is not stored in a matrix, remember that MATLAB provides functions to convert it into one (where possible), such as cell2mat.*

how can you put a single quote character in your string without having it mean the end of the string? Try this command in the Command Window:

>> *fprintf ('This isn't a good place for a quote');*

The fprintf function sees that as the string 'This isn' followed by some illegal text where it expected a comma, plus an unmatched opening quote mark at the end. Errors ensue. Instead, the single quote in the string must be *escaped*. Try this:

>> *fprintf ('This isn''t a good place for a quote');*

The method of escaping a single quote by using two of them is unusual both in MATLAB (shared only by the % sign) and in other programming languages. Normally you escape a character by preceding it with a backslash. Here is a table with the most common escape characters you'll use in strings in MATLAB. Again, more exist and you can see them in the MATLAB Help topic on fprintf.

Tip: *In many programming languages, you must escape the backslashes in a path string, such as "C:\\Users\\ Public". In MATLAB, this is optional when MATLAB expects to see a path.*

Special character	Meaning
' ' (that's two single quotes, not one double quote)	'
% %	%
\n	newline
\t	tab
\\	\

Tip: *To test your output before you send it to a file, set the fileID parameter to 1, which is the ID of the Command Window.*

The format in which you output your data must be known to both you and the recipient (and "the recipient" can mean a coworker, a client, or the rest of the world). This might be a standard to which you must conform or it could be a layout that you've agreed on, including things like column headers. If you both know what to expect, the transfer of information should be problem-free. You have finished Tutorial 8.

Key Terms

path	*absolute path*
working directory	*relative path*

Key Commands

copyfile	fclose	mkdir	save
csvread	fopen	movefile	type
csvwrite	fprintf	pwd	whos
dir	load	range	xlswrite
dlmread	ls	rmdir	

Exercises

Exercise 8.1

Use any function that retrieves a path, then split that path into separate individual folder names using MATLAB string functions.

Exercise 8.2

Assign the results of the path function to a matrix variable. Export that matrix to a plain text file using fprintf. Import the matrix back into MATLAB using the backslash as a delimiter.

Exercise 8.3

Create a script to list the files in your work folder. Assign the list to a variable, and sort the list alphabetically.

PLOTTING AND DATA VISUALIZATION

Introduction

Visualization is a way of exploring ideas, information, and data that lets you tap into different brain processes than that of just looking at lists of numbers. It can make it easy to spot trends, notice outliers, compare different cases, and make observations. MATLAB has a number of ways you can create visual representations of your data. This tutorial introduces you to some of the basic graphing functions.

2D Plots

The command, plot (*X, Y*) creates a 2-D *plot* of X versus Y data. For each Y value the data set must have a corresponding X value. X and Y can be vectors or matrices but they must be equal length. As you may recall, a vector for our purposes in MATLAB is a collection of numbers generally in the form of an array, or single column matrix.

If X and Y are both matrices, they must also be the same size. In this case, the columns of Y are plotted corresponding to the columns of X. Special rules apply in the case where X is a vector and Y is a matrix (or vice-versa). We will talk more about graphing matrices later. For now, we will graph vectors.

Lets try it out by plotting some mathematically derived values. We will start with $y = x/2$. For this type of plot we need actual values (not equations) stored in a single column. You will learn more about graphing symbolic functions in a later tutorial.

Enter the next lines in the Command Window:

>> *x = 1 : 10;*

>> *y = x/2*

Notice the values for y appear as 0.5000　1.0000　1.5000　2.0000 2.5000　3.0000　3.5000　4.0000　4.5000, 5.0000. Just as we expected. Now let's plot it.

>> *plot (x,y)*

The plot appears in a separate window, with its own menu bar across the top similar to Figure 9.1 The *plot window* gives you handy access to many graphing functions. All of the plot types are also available via typed commands which can also be used in scripts or functions.

Objectives

When you have completed this tutorial, you will be able to

1. Use MATLAB to graph functions.

2. Add labels to graphs.

3. Add multiple lines to graphs.

4. Use symbols for graph data.

5. Add a legend to a graph.

6. Create graphs of 3D data.

7. Add a colorbar.

8. Save graphs for use in reports and other documents.

9. Generate code from a graph in the figure window.

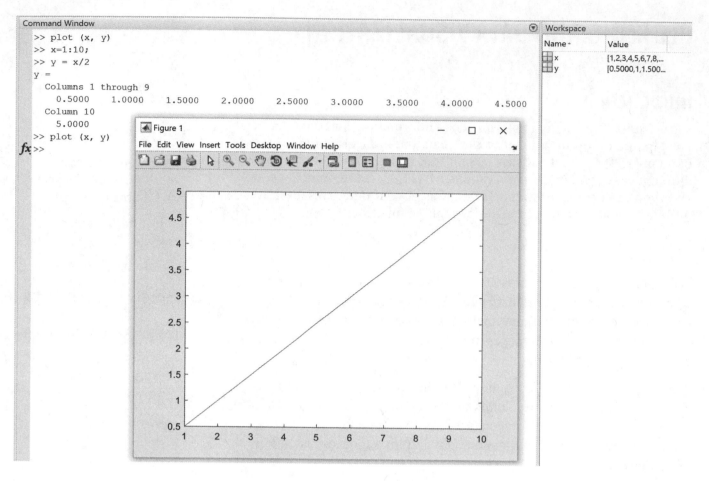

```
Command Window
>> plot (x, y)
>> x=1:10;
>> y = x/2
y =
  Columns 1 through 9
    0.5000    1.0000    1.5000    2.0000    2.5000    3.0000    3.5000    4.0000    4.5000
  Column 10
    5.0000
>> plot (x, y)
fx >>
```

```
Workspace
Name ▴      Value
x           [1,2,3,4,5,6,7,8,...
y           [0.5000,1,1.500...
```

Figure 9.1   The Plot Window

As you would expect, for the values of x (1 through 10) plotted on the *horizontal axis*, the value of y, plotted on the *vertical axis*, is half that of x. Notice that the graph doesn't start at 0,0 as that was not a data point.

> *On your own, create the x vector from 0 to 10 with steps of 1*
>
> *Set y = x/2*
>
> *Plot the graph again to show 0,0 at the origin.*

The new plot replaces the one previously in the plot window. You can click and drag on the borders of the plot window to resize it.

> *Drag the borders to resize the figure similar to Figure 9.2.*

Notice what happens? The *scaling* along the axes expands to the new window proportions. This can be important when you are trying to show the relationships between the data. It is a bit easier to visualize the slope of the line when the axes are at the same scale. Scaling the axes makes it easier to see the values on the scale at times, but may make it harder to interpret what is represented if you can't see the values easily. Keep this in mind when you are making a graph for a slide show for a room full of people, or even when adding data to a report. You can make variations look more or less alarming by scaling the axes. Be nice and represent your data fairly. Be wary when reading graphs when scaling has been used to exaggerate differences in the data.

Tip: *You must set y = x/2 again, because the old variable x contained 10 elements, the new x contains 11 (0 to 10). The two vectors must be the same size.*

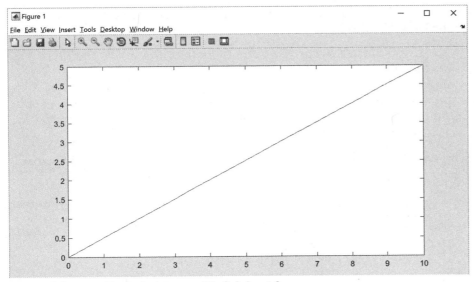

Figure 9.2   **Rescaled Axes with Origin at 0**

The plot window has a number of menu options and handy quick select buttons as shown in Figure 9.3.

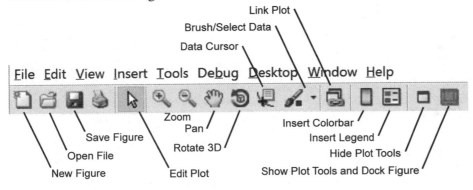

Figure 9.3   **Plot Window Tool Buttons**

 *Click: Show Plot Tools and Dock Figure*

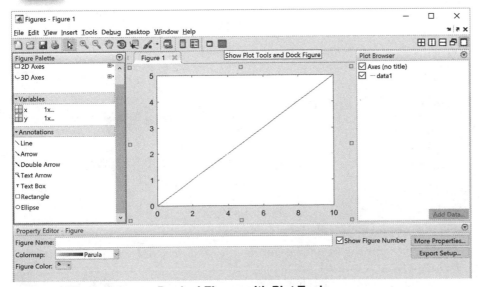

Figure 9.4   **Docked Figure with Plot Tools**

Additional tools appear along the edge of the plot window as shown in Figure 9.4. Next we will explore some of these tools.

Click: **Data Cursor** *from the toolbar at the top of the plot window.*
Move your cursor near the line in the graph. Notice the cursor changes from the selection arrow to a small cross as you approach the data points through which the line is drawn. Click near the central point, where X equals 5.

The data point is identified on the screen. You can click various places along the line to identify the various data points.

Tip: *If your data cursor is not snapping to the data point, you may have previously set this differently than the defaults. If so, see the next step.*

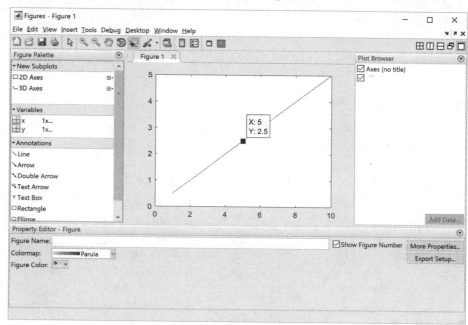

Figure 9.5   **Data Point Identified with the Data Cursor**

Right-click: **in the graph area** *to show the context menu.*

Click: **Selection Style > Mouse Position** *(see Figure 9.6)*

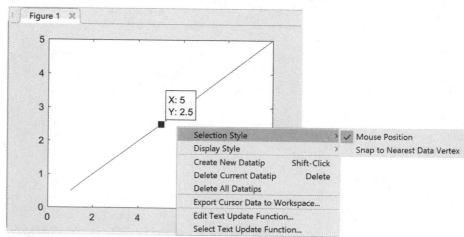

Figure 9.6   **Context Menu for Plot Options**

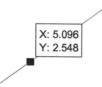

Figure 9.7   Mouse Position on Graph Identified

Move your cursor near the line in the graph. Click near the central point, where X equals 5.

Notice that now, the point selected is on the line, but at the closest mouse position, not on a data point. Either of these features is useful for identifying information from the graph. For now, we will identify the data points.

*Right-click: **in the graph area** to show the context menu.*

*Click: **Selection Style > Snap to Nearest Data Vertex***

Notice the point selected point, jumps back to the closest actual data vertex (and not an interpolated one).

*Click: **Text Arrow** from the toolbars docked to the left of the graph.*

Click and drag to draw an arrow pointing to the data point similar to Figure 9.8

*Type: **Point A** in the input box near the tail of the arrow*

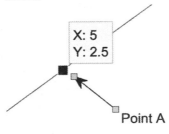

Tip: *If your arrow isn't placed pointing to the data point, click to show its grips. Click and drag a grip to reposition the arrow.*

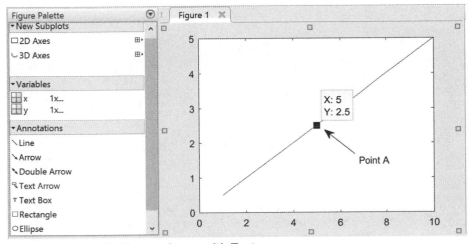

Figure 9.8   **Adding an Arrow with Text**

Adding Titles

It is easy to add titles and labels from the plot window.

*Click: **Axes (no title)** to select it at the right under Plot Browser*

Figure 9.9   **Axes in Plot Browser Area**

The plot window updates to show the property editor for the axes.

On your own, add a title for the plot (such as Fish Caught per Worms Used). Add labels to the X- and Y-axes (Worms Used and Fish Caught, respectively). See Figure 9.10.

Tip: *Notice the tabs near the bottom center of the plot window. Click these to select the X-, Y-, or Z-Axis labels. The Font tab lets you change the label font and size.*

X Axis	Y Axis	Z Axis	Font
X Label:			Worms Use

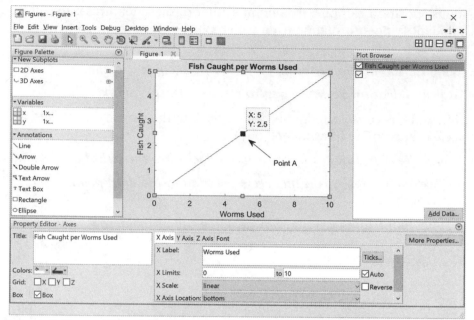

Figure 9.10   **Titles Added to Plot**

Next, lets save this figure to a file.

> Click: **File >SaveAs** *to show the Save As dialog box (Figure 9.11)*

> *Click to expand the options for* **Save as type** *from the bottom of the new window that appear*

You can save the file as a *.fig* type, which makes it easy to reopen and edit in MATLAB. You can also select from a number of popular formats such as *.pdf*, *.jpg*, *.png*, etc. The ability to save to different formats is handy for including your MATLAB graphics in slide shows, reports, and other documents.

On your own, name your file WormsVFish and save it twice. Once as a *.pdf* file and once as a *.fig* file.

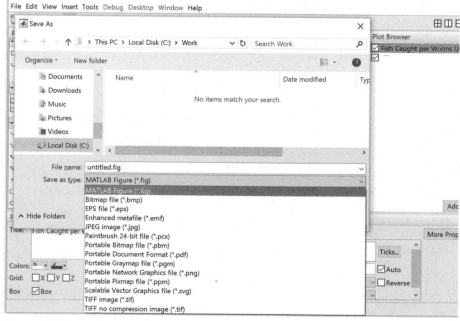

Figure 9.11   **Save As Type Options**

Two new files appear in your file browser, WormsVFish.fig and Worms-VFish.pdf.

Getting Graphics Help

The Help menu from the figure window is a quick way to access help when creating graphics.

Click: **Help > Graphics Help**

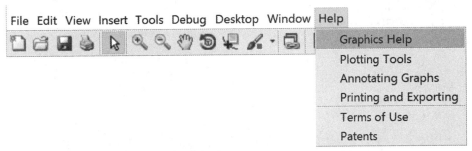

Figure 9.12   **Save As Type Options**

The Help window appears on your screen similar to Figure 9.13 with the documentation for the types of plots. As you can see, MATLAB has a wide variety of plots, charts, and graphs for visualizing your data in both 2D and 3D.

Like plot (x, y), these types of plots may be entered at the command line as well as through the GUI. You can click on the type name to see the many options available.

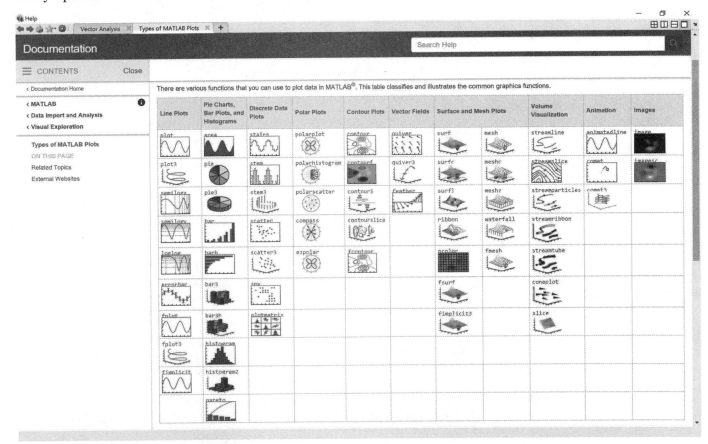

Figure 9.13   **Types of MATLAB Plots**

When you are finished exploring, **close the help window** on your own.

Close the figure window showing your WormsVFish plot.

You return to the MATLAB screen showing the Command window.

Clear the Command Window and Workspace on your own.

Enter the following lines at the prompt.

>> *x = 1 : 10;*

>> *y = sin(x);*

*Click: **PLOTS** tab*

Notice that the buttons are grayed out. This is because no variables are selected.

*Click: **x** from the Workspace list*

*Shift-click: **y** from the Workspace list (hold [Shift] and click y)*

Tip: *The options shown are context sensitive. Not all plot types can be used with every data type. Only the possible options for the selected variables are shown.*
You can click the arrow at the right of the plot types to expand and show even more options when many are available.

Select PLOTS tab

Figure 9.14   **Portion of the Plots Tab**

The buttons on the PLOTS tab are now available.

*Click: **swap arrow** to switch which variable is plotted on which axis*

*Click: **swap arrow again** to switch them back to list x first.*

*Click: **plot** from the buttons on the PLOT tab*

???: *Wait a minute, didn't we expect a nice neat sine wave? But, we only had 10 data points. The plot (x, y) command connects the data points with a line. You will learn more about using symbolic math and the fplot (f) command in a later tutorial. For now, let's add more points to smooth out this plot.*

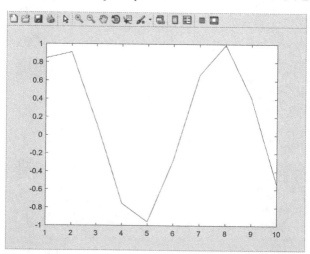

Figure 9.15   **Plot y = sin (x) with 10 Data Points**

*On your own, **close the figure window.***

In the Command Window enter the following:

>> *x = 1 : 0.01: 10;*

>> *y = sin(x);*

*Select **x** and **y** from the Workspace*

Plot x along the x-axis and y along the y-axis on your own.

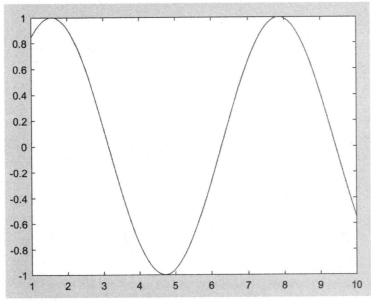

Figure 9.16   **Plot y = sin (x) with More Data Points**

Ah, that's more like it. In the Command Window enter the following:

>> *z = cos(x);*

*Select **x**, **y**, and **z** from the Workspace*

*Click: **Plot as multiple series versus the first input** from the PLOT tab buttons.*

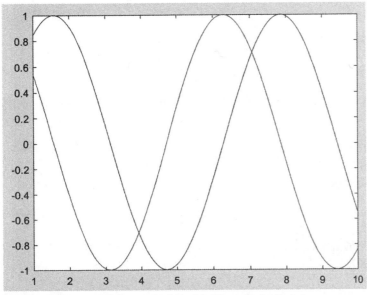

Figure 9.17   **Plot y = sin (x) with More Data Points**

Tip: *You may have to hover your cursor over the icons to see the full name of the plot type.*

Tip: *The Plot as multiple series on the same plot button is very similar. If you see an extra straight line that results from you choosing this option, where x is plotted versus x in addition to y and z.*

Plotting from the Command Line

Notice that when you select the plot button from the tab, the plot command and its options are being run from the command line. This is super handy if you want to include this in your scripts or code. Instead of having to look up each option, you can use the convenient GUI feature and then copy and paste the code into your script or other file. For example, the multiline plot you just created showing the sine and cosine used the following commands, all written into one line:

plot(x,y,'DisplayName','y');hold on;plot(x,z,'DisplayName','z');hold off;

You can also have MATLAB create the code from the figure window.

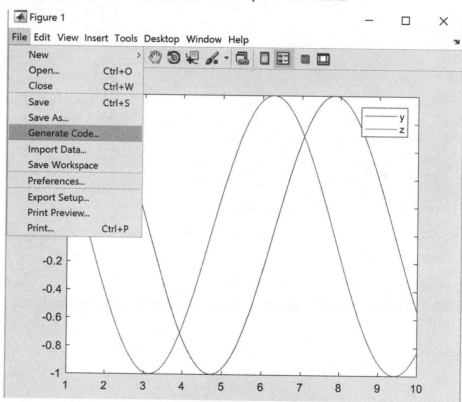

Tip: *The* hold on *command lets you add additional plots instead of overwriting them. Use* hold off *to switch back to where each plot overwrites the last one.*

Click: **Insert Legend** *from the figure window toolbar*

Click: **File > Generate Code...** *from the menu*

Figure 9.18   **Plot y = sin (x) with More Data Points**

The generated code appears in the Editor window similar to that listed here:

Tip: *Your code may not be identical to what is shown here. Some is left out for brevity.*

```
% Create figure
figure1 = figure;

% Create axes
axes1 = axes('Parent',figure1);
hold(axes1,'on');

% Create multiple lines using matrix input to plot
plot1 = plot(X1,YMatrix1);
set(plot1(1),'DisplayName','y');
set(plot1(2),'DisplayName','z');
box(axes1,'on');
% Create legend
legend(axes1,'show');
```

The automatically generated code may not always be optimized. You can always use it as a starting point to make it easy to create your own code.

*On your own, **clear the Command Window and the Workspace***

Charts and Graphs

Bar charts and pie charts make it easy to visualize relationships between data. Both of these can be created easily using MATLAB.

Bar Charts

Bar charts use horizontal or vertical bars to represent categories of data. This makes it quick to visualize quantities, trends, or a change over time. For example a change in population over time, or the variation in kiloWatt usage between day and night or over various years.

Lets give it a try by creating a bar chart from some census data. A spreadsheet file with data for the number (and percentages of households with a computer and access to internet use from the years 1984 to 2012) is provided in the data files.

*Click: **Import Data** from the ribbon Home tab*
*On your own, use the **Current Folder window** to browse to the datafiles.*
*Select to open **2012census-internet-table4.xls***

The Import window shows the data columns.

Make the following changes:
*Output Type: **Column vectors***
*Range: **A8:B19***
Variable Names Row: 7
*Click: **Import Selection***

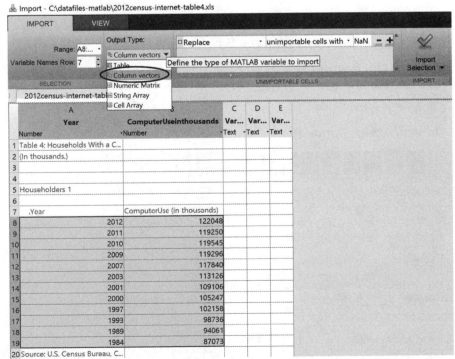

Figure 9.19   **2012 Census Data in Import Window**

The following variables were imported:
Year (12x1)
ComputerUseinthousands (12x1)

The message, "*The following variables were imported: Year (12 x 1), ComputerUseinthousands (12 x 1)*" appears briefly. The data is now imported into the Workspace as a column vector for Years and one for ComputerUseinthousands.

Click: **ComputerUseinthousands** *to select it in the Workspace*

Click: **Bar Graph**

The bar graph appears in the plot window as shown in Figure 9.20.

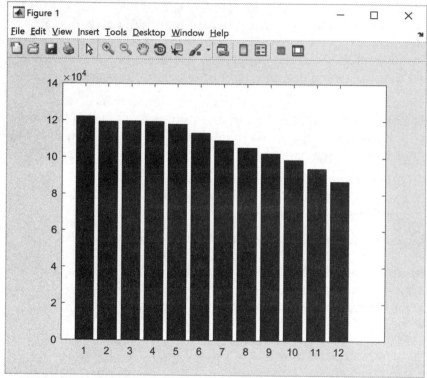

Figure 9.20   **Bar Plot**

Click: **Show Plot Tools and Dock Figure**

Click: **in the graph area**, *notice the border grips appear and the input boxes change*

Click: **Y Axis**

On your own, change the Y Limits to 80000 and 140000

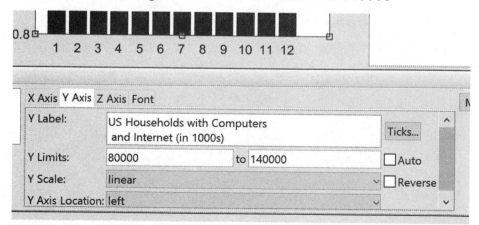

Figure 9.21   **Y Axis Input Boxes**

Add text for the Y Label: US Households with Computers and Internet (in 1000s)

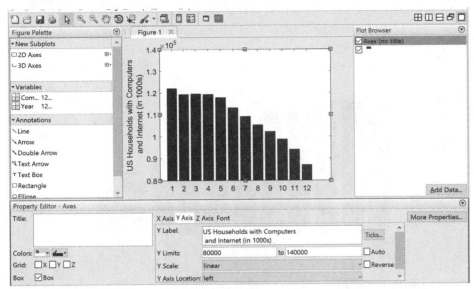

Figure 9.22   Y Axis Label for the Bar Plot

*Click: **X Axis***

*On your own, add **Years** as the **X Label** and change the upper X Limit to **13***

Figure 9.23   X Axis Input Boxes

*Click: **Ticks...** the button to the right of the input boxes*

The Ticks input box lets you automatically or manually enter new tick spacing and labels. Add the years (2012, 2011, 2010. 2009, 2007, 2003, 2001, 2000, 1997, 1993, 1989, 1984) for the items 1-12.

*Click: **Apply***

*Click: **OK***

*Click: **Hide Plot Tools***

The Bar graph looks similar to Figure 9.24.

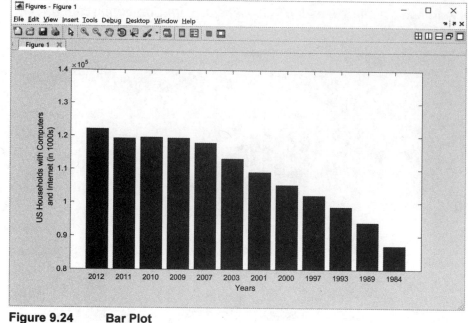

Figure 9.24 **Bar Plot**

*Click: **File >SaveAs** and save your bar chart as InternetBarChart.fig (You may have to change the file type from .pdf to .fig)*

Now let's have a try at using the year data from the Workspace variables.

*On your own, use **clc** to clear the Command Window.*

*From the Workspace, select both **ComputerUseinthousands** and **Year***

If necessary use the swap arrows to list Year first.

*Click: **bar***

The bar plot now automatically uses the year data, but it is scaled to the values as shown in Figure 9.25. There are many ways you can edit the appearance of the chart. You have to decide which way to best represent your data.

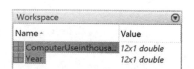

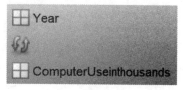

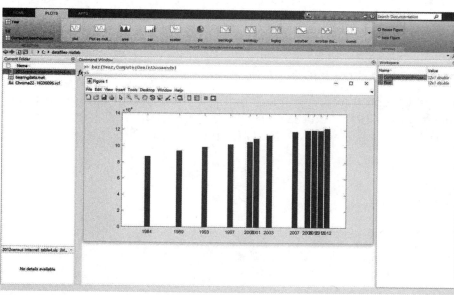

Figure 9.25 **Year Data to Scale in Bar Plot**

On your own, clear the Command Window and the Workspace.

Pie Charts

Pie charts divide a circle into areas representing parts of a whole. They are great for visualizing the relative sizes of categories that make up a whole process, for example, the percentages of the total budget spent on different activities, or the percentages of all returned product based on each type of defect.

*On your own, use **Import Data** to import 25andUp-Education59715. xlsx. Use Output Type: Column vectors. Rename column A to Labels and rename column B to Percentages.*

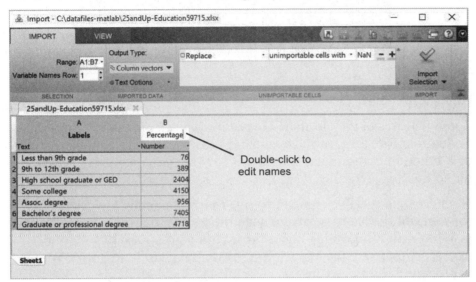

Figure 9.26   Spreadsheet Data Import

Tip: *You can click and drag the borders to resize the columns to read the entire text in the first column.*

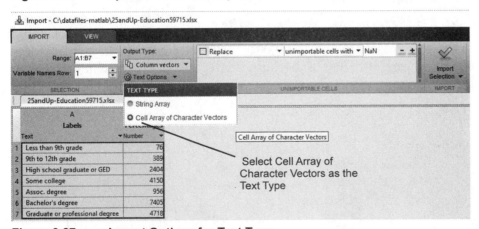

Figure 9.27   Import Options for Text Type

Click: **Text Options**

Click: **Cell Array of Character Vectors** *as the Text Type*

Click: **Import Selection**

The message, "The following variables were imported: Labels (7x1) Percentage (7x1)" appears near the Import Selection button. Notice that now you see Percentage and Labels in the Workspace variables in your main MATLAB screen.

> ## *>> pie (Percentage)*

The pie chart appears similar to Figure 9.28.

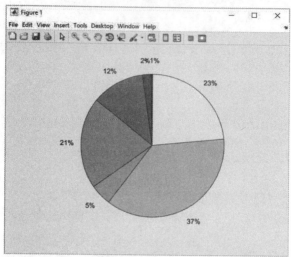

Figure 9.28   Pie Chart

Some features of the pie chart are not available from the GUI interface. So let's explore these at the command line. You can easily make a script or a function using similar statements to those you enter at the command line.

Next create a variable named explode that will contain logical values for slices of the pie to be *exploded* in the plot. In this case you will just enter the data for the logical array, but you can use the relational operators to create the logical values. Our data has 7 categories. We will explode the third one (the logical value 1).

> ## *>> explode = [0 0 1 0 0 0 0]*

> ## *>> pie (Percentages, explode, Labels)*

When the data was imported the text was selected to be a cell array of character vectors. This was because that is the necessary data type for creating automatic labels in the pie chart.

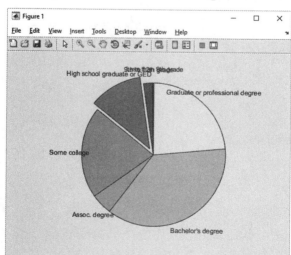

Figure 9.29   Pie Chart with Exploded Section and Labels

Well that didn't look great, because the long label names overlap. It is possible to move their positions, but it will be much easier to just recreate the pie chart without the labels and add a legend to the side. The legend function allows you to specify the location and orientation as well as the labels to use in the legend. We will use eastoutside to place the legend to the right of the pie chart and a vertical list for the orientation. Use help or experiment if you are curious about the other options.

>> *pie (Percentages, explode)*

>> *legend (Labels, 'Location', 'eastoutside', 'Orientation', 'vertical')*

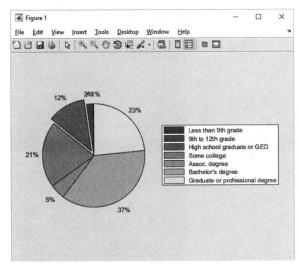

Figure 9.30   Pie Chart with Legend

*Click: **File >SaveAs** and save your pie chart as Education59715Pie.fig*

On your own, close the figure window and clear the Workspace and Command Window.

Plotting a 3D Surface

*On your own, browse to the data files and open **volcano.mat**.*

The file loads and VolcanoSpectralData shows in the Workspace.

*Double-click: **VolcanoSpectralData** to open it in the editor.*

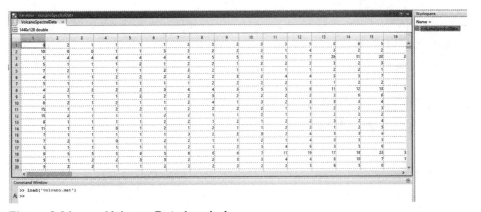

Figure 9.31   Volcano Data Loaded

VolcanoSpectral data is a numerical matrix composed of values from readings from a short-period seismometer on the flank of a volcano in Guatemala. The measurements were taken each minute for 24 hours (1440 values) from each of 128 channels of spectral data from an instrument that is sensitive to ground velocity between 1 and 20Hz. The data has already been filtered and transformed to result the values in the file.

Take a look at the data in the editor, but don't make any changes to it.

*Click the **Plots tab** to show the plotting selections*

*Click: **VolcanoSpectralData** to select it in the Workspace*

*Click: **Surf** from the buttons on the Plot tab*

The 3D surface plot of the data appears in the figure window similar to Figure 9.32.

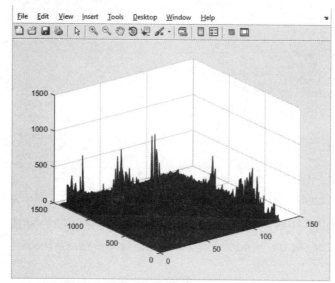

Figure 9.32   **Volcano Data Surf Plot**

*Click: **Mesh** from the buttons on the Plot tab*

The *3D mesh plot* of the data appears in the figure window. Notice how easy it is to try different options and then decide which makes it easiest to visualize your data.

Though the mesh plot looks better than the surface plot, its appearance can still be improved by adjusting the color and the viewing direction.

Changing the Colormap

*Click: **Show Plot Tools and Dock Figure** from the figure window*

The plot tools appear around the borders of your plot.

Click to expand the Colormap selections (see Figure 9.33).

*Click: **Jet** to select it as the color scheme.*

The plot updates to the new colors, but it still does not show enough variation for the volcano data.

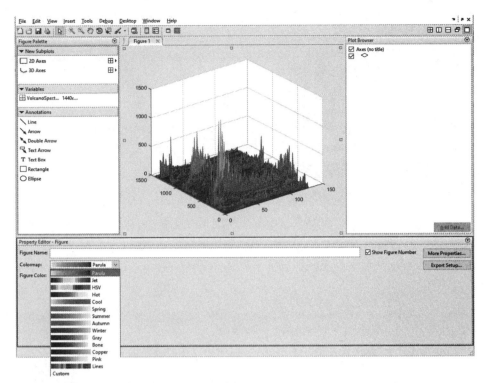

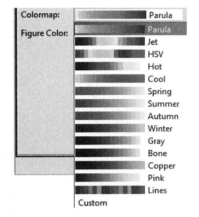

Figure 9.33   Volcano Data Mesh Plot

However, it is very easy to make custom color schemes. Let's try that.

*Click: **Custom***

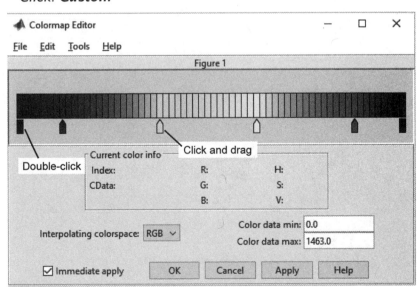

Figure 9.34   Colormap Editor (starting with Jet settings)

The Colormap Editor shows on your screen with the settings from the previously selected color map, in this case that of the Jet colormap. Color markers appear at the bottom of the color bar. You can use them to select new colors and to drag and set the blend from color to color.

Double-click the left most color marker

The Select Marker Color dialog box appears on your screen similar to Figure 9.35.

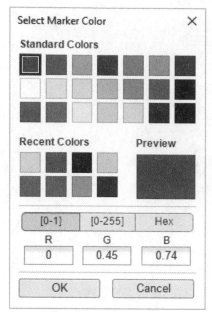

Figure 9.35   Select Marker Color Options

Click on a medium blue color to select it for the color of the left marker

Click: **OK**

Click and drag the markers for turquoise, green, yellow, orange and red colors to the left.

Click: **Apply**

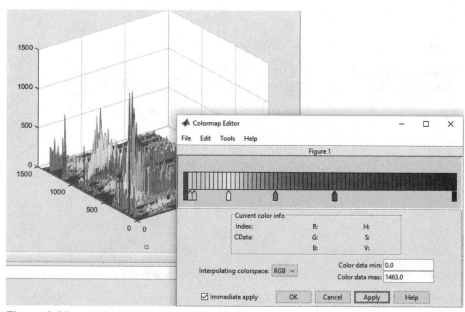

Figure 9.36   Select Marker Color Options

Take a look at the plot in the figure window. Continue to use the Colormap Editor to adjust the colors until you achieve the contrast in the colors needed for visualizing the volcano data. When you are satisfied with the appearance,

Click: **OK**

Changing the 3D Viewpoint

The Rotate 3D selection from the figure window makes it easy to change the 3D viewpoint to produce a better view. This button is a toggle; you click it once to turn the mode on and click again to exit the mode.

 Click: Rotate 3D to select it

With Rotate 3D selected, the cursor appearance changes to a rotation icon.

Click and drag in the plotted figure area to rotate the view so you can see the spikes in the data clearly.

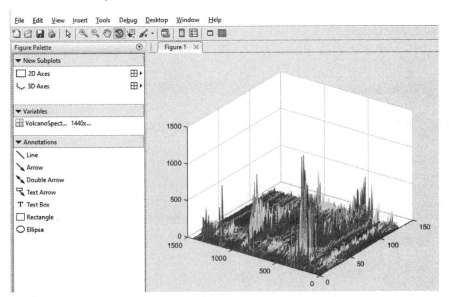

Figure 9.37   View Rotated

Right-click in the plot area to show the pop-up menu

Click: Go to Y-Z view

Figure 9.38   Y-Z View with Pop-Up Menu

Right-click to show the pop-up menu

Click: **Reset to Original View**

On your own, rotate the view to show the data as desired.

Click: **Rotate 3D** *button to deselect it*

Next add a color bar and then axis labels to finish the graph.

Adding a Color Bar

You can quickly add a color bar legend to the graph using the button from the figure window.

Click: **Insert Colorbar** *button*

The color bar appears alongside the plot. It is quite large compared to the graph. You can click and drag to edit it with the selection button active.

Click: **Selection** *button*

Click on the colorbar and drag its corners to resize it

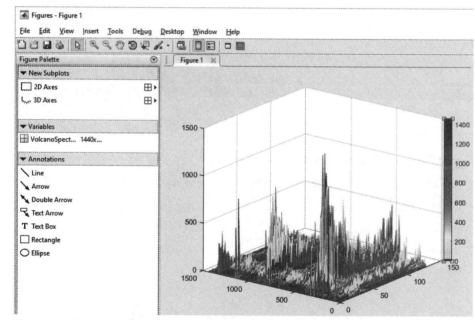

Figure 9.39   Colorbar Resized

Next, finish up the graph by adding a title and axis labels.

On your own, click Axis (no title) from the Plot Browser area at the right of the graph.

Enter the following information for the plot title and axis labels.

> Title: Spectral Data 8/23/2015
>
> X Label: Station Number
>
> Y Label: Measurement Time (minutes)
>
> Z Label: Spectral Reading

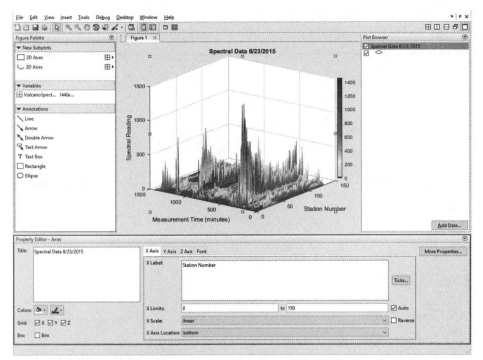

Figure 9.40   View Rotated

Use the selection arrow and click the axis labels and drag them to reposition them.

Use File, Save to save your file as a MATLAB figure. Use SaveAs to save a .jpg or other format.

Select File, Generate Code and view the code in the MATLAB Command Window Editor.

The beauty of the GUI plotting tools is the ease of exploring and producing a good-looking graph. From that you can generate the code and then automatically produce a series of similar graphs. Notice also the commands are issued in the MATLAB Command Window for each selection and change you make in the figure window. You can show the command history and edit it to get started producing the code needed for plots.

This volcano spectral data is measured daily, so automating the process to visualize the data can be a real time saver. Plus notice how easy it is to identify the significant events recorded in the data when they are graphed this way compared to examining rows of numbers.

There is lots more to explore in the plotting functions, but hopefully this introduction will provide the exposure for you to explore it more on your own.

Congrats! You have finished Tutorial 9.

Key Terms

.fig	*bar charts*	*pie charts*	*vertical axis*
2-D plot	*exploded pie chart*	*plot window*	
3D plot	*horizontal axis*	*scaling*	

Key Commands

hold off   plot (x, y)
hold on

Exercises

Exercise 9.1
Graph the equation $y = x^2$ from x = -3 to x = +3. Use a step of .1 between values in that interval.

Exercise 9.2
Create a graph of $y = 2x - x^2$ over the interval x = -2 to x = + 4. Determine what step to use to show a smooth curve.

Exercise 9.3
Create a graph of y = sqrt (x - 1).

Exercise 9.4
Create a graph of $y = x^3 - 12x$.

Exercise 9.4
Create a graph of the mortality data for Finland for the years in the data set provided in *Mortality-1.xlsx*

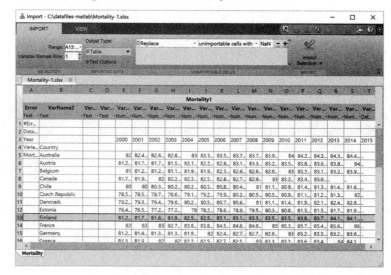

MATLAB TOOLBOXES: CURVE FITTING

Introduction

As you have seen so far, MATLAB has a wide array of features. This is extended further through add-on *toolboxes* for a wide range of applications and features. The typical student suite for MATLAB includes Simulink, and the following toolboxes: Control Systems, Data Acquisition, DSP System, Image Processing, Instrument Control, Optimization, Signal Processing, Simulink Control Design, Statistics & Machine Learning, and Symbolic Math. The toolboxes are add-ons to the main MATLAB software, so you may or may not have them available.

The main help screen page shows the toolboxes you have available.

Click: Help

Figure 10.1   Help Screen Showing My Products

The left side of the help screen shows the products for which you have licenses. You may see a different list than shown here. If you do not have the Curve Fitting Toolbox licensed, you will not be able to complete this tutorial. You can purchase add-on toolboxes at any time if you so desire.

Apps Tab

The Apps tab (short for Applications) provides a GUI interface to the your add-on toolboxes. You can use this tab to browse available apps and purchase them (via the Get More Apps button), install additional apps and toolboxes (Install App), and even package your own apps to make them available to others (Package App).

Objectives

When you have completed this tutorial, you will be exposed to:

1. MATLAB toolboxes.
2. Curve Fitting toolbox.
3. Basic issues in curve fitting.

Click: **Apps tab**

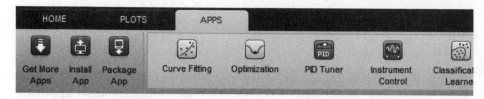

Figure 10.2   The Apps Tab

Your Apps tab may have different buttons to the right depending on which applications you have licensed and installed.

Curve Fitting

Curve fitting is the process of generating an equation or function from given data points. This equation can then be used to predict or extrapolate results beyond that of the measured data.

For example, a school district may measure enrollment and other data over a number of years and use that data to generate an equation letting them predict the number of students in future years.

Let's look at an example using the computer use data file stored in the file *ComputerUse2012.mat*. This file has years sampled from 1989 through 2012 and the number of households in thousands using computers.

> *On your own, browse to the data files and load ComputerUse2012.mat* **into MATLAB**

You should now see two variables in the Workspace, *Year*, and *ComputerUseinthousands*.

Lets do a quick scatter plot of the data to review it.

> *On your own, use the PLOTS tab and select variables Year and ComputerUseinthousands and select Scatter*

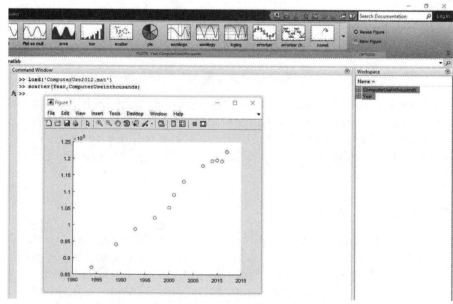

Figure 10.3   Scatter Plot of Year v. Computer Use

Figure 10.3 shows the scatter plot of the year versus computer use data. Next let's fit a curve to this data.

Click: **Apps tab**

Click: **Curve Fitting**

The Curve Fitting Tool opens on your screen. Use it to make the following selections:

X data: **Year**

Y data: **ComputerUseinthousands**

Select **Polynomial** in the central area for the fit type.

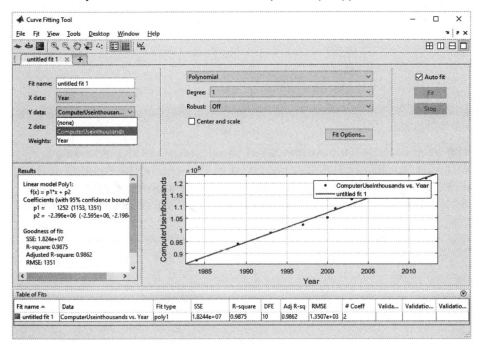

Figure 10.4   Curve Fitting Tool

Examine the upper portion of the Results panel of the Curve Fitting Tool near the bottom left of the window.

```
Linear model Poly1:
    f(x) = p1*x + p2
Coefficients (with 95% confidence bounds):
    p1 =       1252  (1153, 1351)
    p2 = -2.396e+06  (-2.595e+06, -2.198e+06)
```

This area shows the fit equation form ($f(x)=p1*x + p2$) followed by the specific coefficients for the curve for your data ($p1 = 1252$ and $p2 = -2.396e+06$). So, for the first degree polynomial fit, the equation for your data would be:

$$f(x) = 1252x - 2396000$$

MATLAB provides some analysis for how well the fit curve matches the data points in the lower portion of the Results panel under the heading Goodness of Fit.

Tip: *You will learn more about polynomial order in this tutorial. You can also check out Wikipedia.*

Goodness of Fit

This information in the Results area of the Curve Fitting Tool lets you know how well the curve equation matches the data points. You can use the statistics available here to compare the fits you get using the various fit algorithms. Statistics reporting a smaller range of error between the result from the fitting equations and the data points helps indicate that the equation may result in a more accurate prediction. Detailed explanations of analysis of fitting statistics would be a course in itself, so only brief descriptions are presented here. You can read more about these topics in the MATLAB help files.

For the computer use data result the fit statistics reported were:

> Goodness of fit:
> SSE: 1.824e+07
> R-square: 0.9875
> Adjusted R-square: 0.9862
> RMSE: 1351

Sum of Squares (SSE)

The *SSE value* is a measurement of the deviation of the actual data values from the predicted curve given as the sum of the squares of these deviations. The sum of these squares is always a positive number. The closer this value is to zero, the better the fitted curve predicts the data.

Tip: *You can click and drag the legend to a better location. Use the corner grips to resize the legend when necessary.*

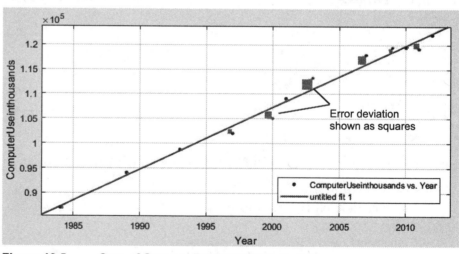

Figure 10.5   **Sum of Squares Deviations Illustrated**

R-square

The *R-square value* is a ratio used to correlate the actual data compared to the predicted values based on the fitted curve. This value ranges from 0 to 1 (though it can be a negative number, but this indicates severe problems with the fit.) A value of 1 indicates 100% correlation between the data and the fit curve. A value of .9875 indicates the fitted curve explains the 98.75% of the variation in the data.

A value closer to 1 indicates the curve fits the data better, but does not indicate that this curve represents a correct model for predicting future values. And in some fields like psychology or medicine, that predict human behavior, lower R-square values may be acceptable as human behavior is harder to predict than that of mechanical systems.

Making changes to the fit equations to achieve a better R-square value may not improve the accuracy of predictions based on that equation as many factors affect the accuracy of a model for making predictions. In the case of our computer use data, many factors are at play in deciding how many households in the US use computers, such as the price of computer technology, population growth, cell phone use supplanting computers, availability of high speed internet, market saturation, and other factors. For the computer use data the small number of data points should also be a huge concern.

Adjusted R-square

This measure refines the R-square statistic based on the number of predictors in the model. The range of the value is similar to that of R-square.

Root Mean Squared Error (RSME)

The RMSE value estimates the error by measuring the difference between the actual values and the predicted values (called the residuals) and then squaring the residuals, averaging those squares and finally using the square root of that average. A value closer to 0 indicates a better fit.

You can plot the residuals as an aid in visualizing the fit. A quick way to do this is to use the Residuals plot button found on the Curve Fitting Tool menu. You cannot see the menu bar if the Curve Fitting Tool window is docked in your MATLAB Command Window area.

Click: Residuals plot button

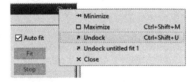

TIP: *If you do not see the menu bar, press CTRL + Shift + U or click the small arrow at the upper right of the window and choose Undock.*

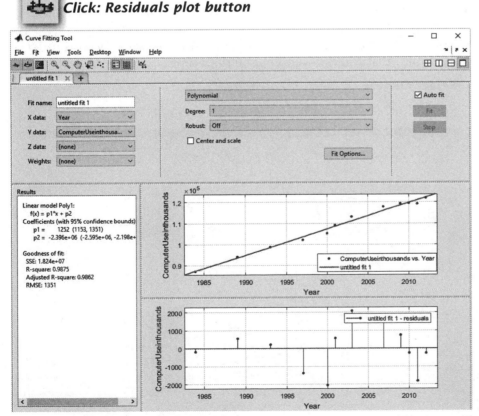

Figure 10.6   Residuals Plot

Fitting Methods

Now that we have taken a brief look at goodness of fit, let's explore the different methods MATLAB has available for generating the fit curve. Fit curves are often used as the basis for making predictions based on measured data, so understanding curve selection is another important part of the process. This is just an introduction to curve fitting. In order to make good predictive models you will need to study much more about statistical analysis.

Polynomial Order

We used *polynomial* as the curve type for the computer use data. As you may recall from your math classes, a polynomial equation is one of the general form: $f(x) = a_n x^n + a_{n-1} x^{n-1} + \ldots + a_2 x^2 + a_1 x + a_0$. This is essentially a coefficient (a) times a variable (x) raised to a power (n). The degree of the polynomial is the highest exponential power to which the variable is raised. For example, $f(x) = 7x^3 + 5x^2 + 2$ is a third degree polynomial as the highest power of x is 3.

In general, when you graph a polynomial function, there is one less possible inflection point in the graph than the degree of the polynomial. A first order equation such as $f(x) = 3x + 2$ or the first order equation for our computer use data is a straight line.

An equation like $f(x) = x^2$ has one inflection point and is a parabola. An equation such as $f(x) = x^3$ has up to two inflection points. A 4th order polynomial may have up to three inflection points, though it is possible for it to have fewer depending on all of the terms in the equation, and so on for each higher order of polynomial.

What does this mean for curve fitting? A very high order polynomial may pass through more points in your data than a lower order one. This sounds good at first, but you should generally choose the lowest order polynomial that still fits your general data trend.

Select Degree : 7 (below Polynomial)

Click: Center and scale (so that it appears checked)

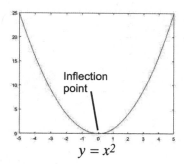

$y = x^2$

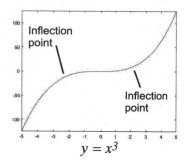

$y = x^3$

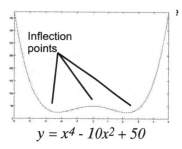

$y = x^4 - 10x^2 + 50$

Tip: *You may see a warning that the curve is badly conditioned. Clicking Center and Scale should resolve this. However, this warning is generally information you would not ignore.*

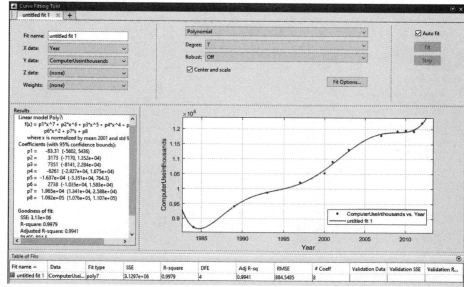

Figure 10.7 **Exponential Growth Fit for Computer Use Data**

Now the R-square value is 0.9979! Does this mean that this curve fits the data best? Well, perhaps the answer is yes, if we mean does it match to the data points best, but NO if we are asking is it the most useful for making predictions based on the equation. This fitting equation is:

$$f(x) = -83.31x^7 + 3173x^6 + 7351x^5 - 6261x^4 - 16370x^3 + 2738x^2 + 19650x + 109200$$

This is probably not a good choice, though it does give a high correlation between the fit curve and the data.

Exponential Growth

Click to expand the fitting options from the central panel in the Curve Fitting Tool

On your own, scroll down and examine the full list.

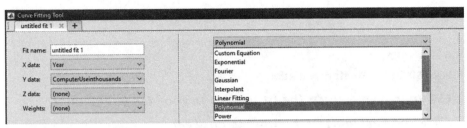

Figure 10.8   **Fitting Types Expanded**

*Click: **Exponential** from the types available*

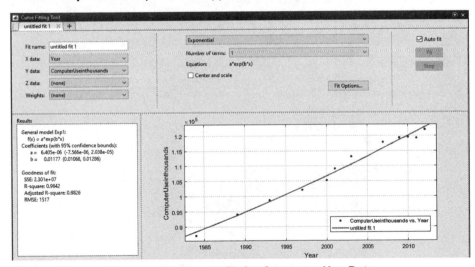

Figure 10.9   **Exponential Growth Fit for Computer Use Data**

Notice that the fit is still pretty good in a statistical sense. The R-square value is 0.9842. But it reflects a belief that the rate of change in households with computers will continue to increase, when in fact, we would expect market saturation to cause the rate to slow over time.

This curve makes sense for certain types of data, but again, you should take care when using this for predictions.

Custom Equation

You can enter a custom equation as the general type for use as the fitting equation using the Custom selection. This is very useful when you know the general form of equation that should fit your data.

Custom Equation

$$y \quad = f(\; x \quad)$$
$$= \; 1\; a*x+c$$

Tip: *A good overview of curve fitting in general is available at http://physics.info/curve-fitting/.*

Select Custom *(from the fit type drop down list)*

Change the formula in the input box to *a*x+c*

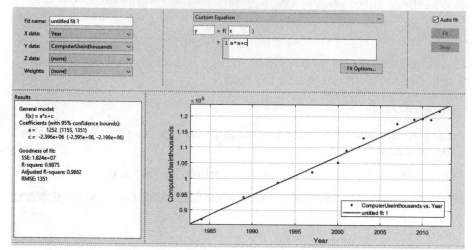

Figure 10.10   Custom Equation

On your own, return the fit type to first order polynomial.

Click the **Residuals plot** button to unselect it.

Click: **Tools > Prediction Bounds > 90%**

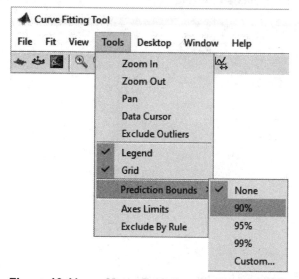

Figure 10.11   Menu Selection for Prediction Bounds

The plot now shows dashed lines around the range within 90% confidence bounds similar to Figure 10.12.

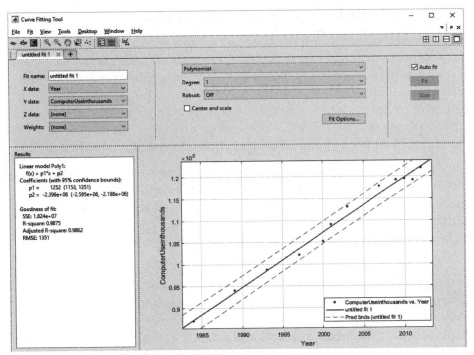

Figure 10.12   90% Confidence Boundary

Sample Size

If you look at our computer use data, we only have 12 data points in the sample. If any of the data years are *outliers* (exceptions to the norm or even incorrectly measured data) then those points will tend to "drag" your fit curve in their direction. This is particularly problematic when the outlier is near the end of the fit curve, as it will make your projections based on the curve off even more toward the end of the data range, where you would be most interesting in forecasting a value.

Lets try our fit equation to make a projection for 2016 (a census year which will soon have published data). Using the coefficients from the results pane, the first degree polynomial fit equation was:

$$f(x) = 1252x - 2396000$$

The independent variable, x is the year, so projecting for 2015 we get:

$$f(x) = 1252 * 2018 - 2396000$$

$$= 130536 \text{ (x 1000 as the data was in thousands)}$$

$$= 130,536,000 \text{ computers in use projected in 2016 compared} \\ \text{to } 122,048,000 \text{ in 2012 from the census data.}$$

Once the 2016 census data is published, you can check to see if this prediction is close to the recorded data for that year. If the prediction is not correct, what factors can you take into consideration to improve this model?

Tip: *From the Curve Fitting Tool window, you can select File > Generate Code to show the commands needed to produce the fitting curve.*

```
%% Fit: 'untitled fit 1'.

[xData, yData] =
prepareCurveData (Year,
ComputerUseinthousands);

% Set up fittype and options.

ft = fittype( 'poly1' );

% Fit model to data.

[fitresult, gof] = fit( xData,
yData, ft );

% Plot fit with data.

figure( 'Name', 'untitled fit 1' );

h = plot( fitresult, xData, yData );

legend( h,
'ComputerUseinthousands vs.
Year', 'untitled fit 1', 'Location',
'NorthEast' );

% Label axes

xlabel Year

ylabel
ComputerUseinthousands

grid on
```

Heat-Treatment Curve Fitting Example

Heat-treated materials are stronger and more resistant to bending and deterioration due to abrasion and rust. The treating routine involves sequential immersion in heating and quenching environments following a strict schedule dictated by the material's internal physics. Heating for a shorter or longer period can make the heat-treating totally ineffective causing these parts to fail when used. What matters is not just the temperatures at which you anneal or quench, but how fast or slowly it is done.

Heat-treating depends critically on understanding what these temperature measurements tell regarding the heating and cooling rates, and whether these rates are outside the allowed limits. Our goal is to find the representative function for the temperature over time. Later we will use this to find the heating or cooling rate, which is the slope (derivative) of that continuous algebraic function based on the fit curve.

If the points are accurate measurements of time and temperature, then the function should pass through every data point. It is often the case that data points are "scattered" and only approximate (due to measurement errors, instrument problems and other reasons) so at best the curve gives a realistic trend indicated by the points by fitting a curve through the vicinity of the points with a minimum statistical deviation from them.

For our data we want to find the likely variation of temperature in the furnace. We also want to see if the part is an acceptable one based on the observed heating and cooling rate data. The process engineer requires the manufactured part be rejected if the heating rate exceeded 600° F per hour or if the cooling rate exceeded 400° F per hour. The heat-treating process lasts 12 hours and the temperature ranges from 80 to 2600° F. In the data file, the temperature (T) points are designated in kiloFahrenheit to reduce the number of digits, and the time (t) in decimal hours. For example, a temperature point (t=3.95, T=0.811) means that 3.95 hours (3 hours and 57 minutes) into the process the temperature was 811° F.

Time (Hours after start)	Temperature (in 1000 of degrees F)
1.0	0.101
2.0	0.466
3.0	1.03
4.0	1.8
5.0	2.22
6.0	2.04
7.0	2.3
3.5	0.9
4.2	1.5
8.0	1.8
10.0	1.99
11.0	2.8
12.0	2.406

We want to find a suitable equation from the curve fit for the recorded temperature points and use it to determine the heating or cooling rate.

We will use a polynomial fit of the temperature data. Keep in mind that there is no absolute rule for which degree polynomial is "best". The higher the degree, the more likely the curve will hit every data point (the smaller the rms deviation). However, the penalty is often that the slope of the curve will vary drastically. In the case of heat-treating that will mean artificially large heating/cooling rates will be determined which will cause rejection of manufactured parts that are acceptably heat-treated.

On your own, import the spreadsheet HeatTreatData.xlsx provided with the data files.

Import the Time and Temperature data as Column Vectors.

Create a scatter plot similar to that in Figure 10.13.

Tip: *A rule of thumb is that the maximum degree of the polynomial should be one unit smaller than the number of points available for fitting, e.g. if there are seven points available, the polynomial degree should be at most six.*

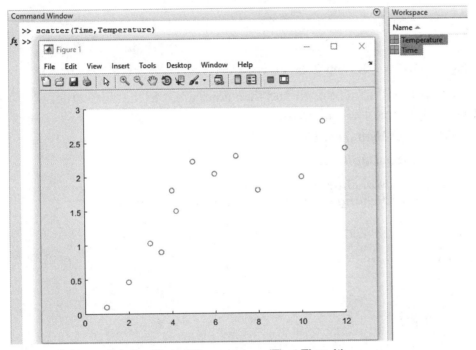

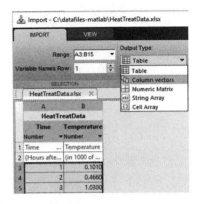

Figure 10.13 **Scatter Plot of Temperature (T) at Time (t)**

Tip: *If your plot is not similar to the figure, you may need to use the Swap button to switch the variable order.*

Now use the Curve Fitting app with Time as the X Data and Temperature as the Y Data.

Change the Degree to 2 for a second order polynomial fit.

Your results should be similar to Figure 10.14.

    (x) = p1*x^2 + p2*x + p3
    Coefficients (with 95% confidence bounds):
        p1 =   -0.02871 (-0.04959, -0.007824)
        p2 =   0.5729 (0.2892, 0.8566)
        p3 =   -0.4151 (-1.211, 0.3807)
    Goodness of fit:
     SSE: 1.181
     R-square: 0.847
     Adjusted R-square: 0.8165
     RMSE: 0.3437

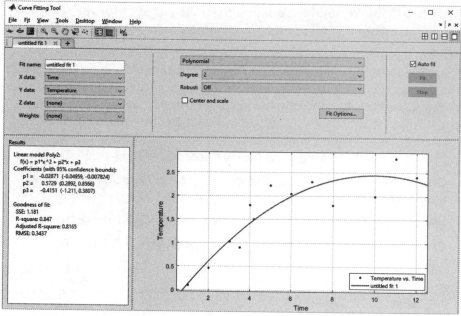

Figure 10.14   **Second Order Polynomial Fit of Temperature (T) at Time (t)**

Next you will name this fit and then compare it to the third order polynomial fit that you will create next.

Fit Name: **Heat Treat Poly 2**

Click: **Fit > Duplicate Heat Treat Poly 2**

On your own, change the name for the copied fit to Heat Treat Poly 3 and change the Polynomial order to degree 3.

Tip: *If your screen does not look like the figure, use Window, Maximize to show each curve in overlapping taps. Experiment with the Left/Right tile and Top/Bottom tiles on your own.*

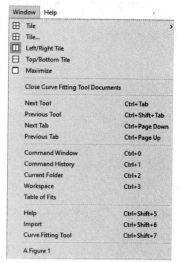

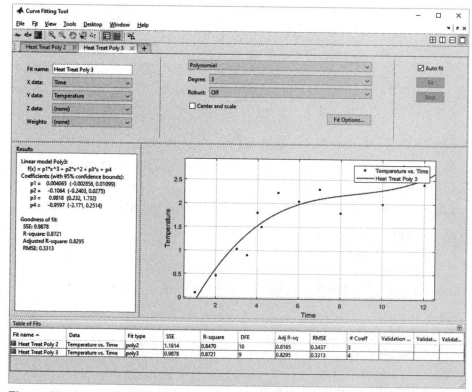

Figure 10.15   **Third Order Polynomial Fit of Temperature (T) at Time (t)**

Now you have two tabs and the Table of Fits at the bottom of the screen shows both sets of fit data similar to Figure 10.15.

For the third order polynomial fit, the equation coefficients are:
> Linear model Poly3:
> > $f(x) = p1*x^3 + p2*x^2 + p3*x + p4$
>
> Coefficients (with 95% confidence bounds):
> > p1 =   0.004065 (-0.002858, 0.01099)
> > p2 =   -0.1064 (-0.2403, 0.0275)
> > p3 =   0.9818 (0.232, 1.732)
> > p4 =   -0.9597 (-2.171, 0.2514)

Notice the R-square value has improved and visually, this 3rd order fit appears to match the points better too.

> ***Own your own, add the fits for the fourth, fifth, and sixth order polynomials.***

The curve fits for the 2nd to 6th polynomial orders are now added. The Fit Table at the bottom of the screen lets you easily compare the fit values for each of the fits, see Figure 10.16.

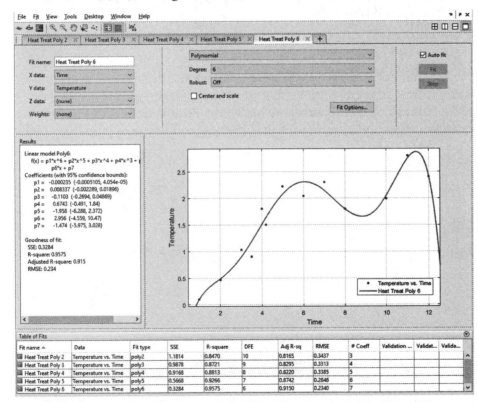

Figure 10.16   **Sixth Order Polynomial Fit of Temperature (T) at Time (t)**

You have the data fit with a series of polynomials of increasing degree, from 2nd to 6th as follows:

$f(x) = p1*x^2 + p2*x + p3$
$f(x) = p1*x^3 + p2*x^2 + p3*x + p4$
$f(x) = p1*x^4 + p2*x^3 + p3*x^2 + p4*x + p5$
$f(x) = p1*x^5 + p2*x^4 + p3*x^3 + p4*x^2 + p5*x + p6$
$f(x) = p1*x^6 + p2*x^5 + p3*x^4 + p4*x^3 + p5*x^2 + p6*x + p7$

Let's check out how well the equations project our temperature data.

Next you will use the MATLAB Command Window to create a new script file where you will copy and paste the equations and coefficients from the Curve Fitting app.

Click: **to activate the Command Window**

>> Edit curve4

On your own select Yes to create the file *curve4.m.*

The Editor opens with curve4.m as the file name.

On your own, navigate to the 4th degree polynomial fit in the Curve Fitting app.

Highlight the equation and coefficients in the Results area for the 4th degree polynomial fit and copy them to the clipboard using either right-click and select Copy or pressing Ctrl+C.

Navigate to the MATLAB Editor window for curve4.m and paste the clipboard contents into the Editor window as shown in Figure 10.17. You can either right-click and select Paste, or press Ctrl+V to do this.

On your own, edit the text you pasted from the clipboard as shown in Figure 10.17.

Notice that you will add a value for x before the equation. This is the value to test to see how the curve predicts the temperature value. The extra values in parentheses have also been removed. The f(x) has been changed to y4 and comment % has been added to the explanatory text.

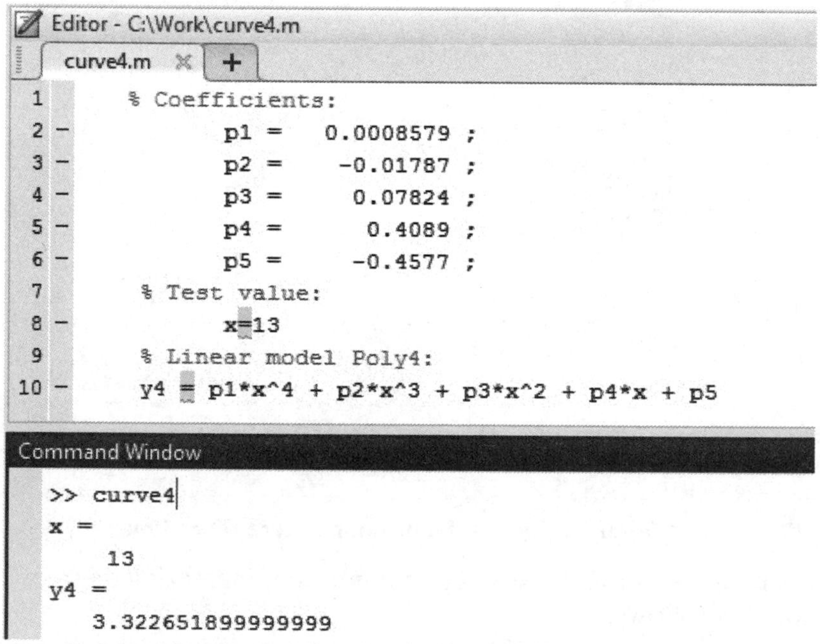

```
Editor - C:\Work\curve4.m
curve4.m  ×  +
1        % Coefficients:
2  -          p1 =    0.0008579 ;
3  -          p2 =     -0.01787 ;
4  -          p3 =      0.07824 ;
5  -          p4 =       0.4089 ;
6  -          p5 =      -0.4577 ;
7        % Test value:
8  -          x=13
9        % Linear model Poly4:
10 -      y4 = p1*x^4 + p2*x^3 + p3*x^2 + p4*x + p5
```

```
Command Window
>> curve4
x =
     13
y4 =
    3.322651899999999
```

Figure 10.17   Equation with 4th Order Coefficients in the Editor

Tip: You can show more or fewer decimals in the Command Window using the format command.

Click: **to activate the Command Window**

>> curve4

So for x = 13, y4 = 3.322651899999999.

On your own, use the Editor to change the value for x to 5.

Run curve4 in the Command Window again.

The result for x = 5 is y4 = 1.845237500000000.

Try it again on your own, with x = 1.

For x = 1, the result is y4 = 0.012427900000000.

How do these values stack up compared to the actual data points? We did not have a value provided for x = 13, but for x = 1 and x = 5 the values from the data were as listed:

Time (Hours after start)	Temperature (in 1000 of degrees F)
1.0	0.101
5.0	2.22

On your own, start a new script file curve5.m in the Editor.

Return to the Curve Fitting App window and copy and paste the coefficient results and formula into the Editor window.

Edit the pasted information to be similar to that shown in Figure 10.18, with x = 13.

Run the curve5.m script in the Command Window.

```
Editor - C:\Work\curve5.m
  curve4.m      curve5.m      +
1        % Coefficients:
2              p1 =   -0.0007062 ;
3              p2 =      0.02394 ;
4              p3 =      -0.2925 ;
5              p4 =         1.52 ;
6              p5 =       -2.774 ;
7              p6 =        1.703 ;
8        % Test value:
9              x=13
10       % Linear model Poly5:
11             y5 = p1*x^5 + p2*x^4 + p3*x^3 + p4*x^2 + p5*x + p6
12
```

```
Command Window
>> curve5
x =
    13
y5 =
    1.441723399999984
```

Figure 10.18   Equation with 5th Order Coefficients in the Editor

The result for x = 13 is y5 = 1.441723399999984. Compare this to the curve4 result of y4 = 3.322651899999999. Keep in mind that the data is in thousands of degrees, so this difference is very significant when it comes to making predictions. Try the curve5 script with x = 1 then x = 5.

x = 1, y5 = 0.179733800000000
x = 5, y5 = 2.026125000000001

So thinking about which of these curves will better represent the data trend is not a cut and dried process, particularly when there are not many data points. One way to improve the result is to collect more data!

In our heat treatment example, we are actually interested in the rate of change in the heating and cooling process, in other words, we want to find the first derivative of the equations which will show us the slope, or rate of change. These rate curves formed by differentiating the temperature curves, will give us a better picture. By investigating them, we may gain insight into which of the curve fitting equations best represents our heat-treatment scenario.

To find the first derivative of the temperature curve we could use basic differential calculus. For example, if the temperature curve is described by the fourth-degree polynomial as follows:

$$f(x) = p1*x^4 + p2*x^3 + p3*x^2 + p4*x + p5$$

then the rate curve will be the third-degree polynomial,

$$f'(x) = 4*p1*x^3 + 3*p2*x^2 + 2*p3*x + p4$$

By plotting the rate curves we will get a better picture for the restrictions on the heating or cooling rates allowed. Well, coincidentally enough, using symbolic math for derivatives and plotting symbolic functions is the topic for the next chapter, so we will return to investigating this heat-treating data question there.

Exit the Matlab software for now as you have completed this tutorial!

Key Terms

curve fitting	*R-square value*	*toolboxes*
polynomial order	*SSE value*	

Key Commands

Exercises

Using the data file, *USPopulationData.xlsx* where the population of the USA in 1920 was 106,461,000 and the population in 2010 was 310,232,863 to answer the following questions:

Exercises 10.1
Projecting exponential growth, what will be the projected population of the USA in 2020?

Exercises 10.2
Write the curve fit equation for the exponential growth model for the data. In your opinion is an exponential growth model the best projection? Present at least one alternative curve fit equation.

Exercises 10.3
The population recorded for 2016 was 324,118,787. What does your exponential growth model predict? What is the difference between the collected data and that of your model?

Exercises 10.4
What is the projected population of the USA in 2043?

Exercise 10.5
Given the recorded heat-treatment data from the file, *HeatTreat-Data.xlsx*, find the likely variation of temperature in the furnace.

SYMBOLIC MATH

Introduction

MATLAB's *Symbolic Math Toolbox* lets you manipulate and solve mathematical equations symbolically. Up to this point, variables you used in equations were assigned a numeric value of some sort before they could be used in equations. Using the Symbolic Math Toolbox, you can define symbolic variables and use them in equations in a way that is very similar to solving mathematical equations as you would by hand. You can use these features for calculus, linear algebra, ordinary differential equations, simplifying equations, and other functions. You can also plot symbolic equations. Using the *Live Editor* you can add explanation and show standard mathematical typesetting notation for equations. This is useful for preparing papers and other documentation.

Getting Started with Sym Variables

Up to this point, you assigned a value to a variable before using it. The syms object type lets you use variables symbolically.

The fplot function lets you plot a curve defined by a function. With fplot you can specify an interval, but if none is specified, the default range of -5 to +5 is used when there is no interval. Try it out with the following commands:

```
>> syms x
>> f(x) = sin (x)
>> fplot (f)
```

The plot window appears similar to Figure 11.1.

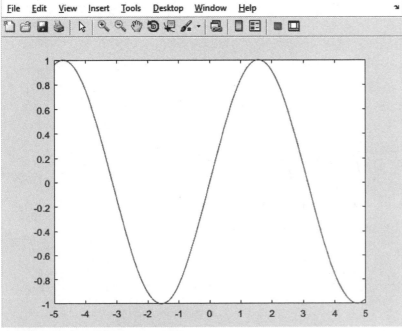

Figure 11.1   Plot of f(x) = sin x

Objectives

When you have completed this tutorial, you will be able to

1. Use the Live Editor to add explanation and formatting to your MATLAB project.

2. Create symbolic variables.

3. Use symbolic mathematical expressions in MATLAB.

Line parameters can be listed in any order. Defaults are used for any you do not specify.

Line pattern symbols	
-	Solid (default)
--	Dashed
:	Dotted
-.	Dash-dot
Marker symbols	
o	Circle
+	Plus sign
*	Asterisk
.	Point
x	Cross
s	Square
d	Diamond
^	Upward triangle
v	Downward triangle
>	Right triangle
<	Left triangle
p	Pentagram
h	Hexagram
Color symbols	
y	yellow
m	magenta
c	cyan
r	red
g	green
b	blue
w	white
k	black

In a previous tutorial you used **plot** to show a similar curve, but in that case, specific values had to be assigned to the variable before it could be used. With symbolic variables you can explore equations and using fplot, you can visualize symbolic functions. Let's check out a few more. This time we will use some fplot features to change the line pattern and color. The form **fplot(arguments, 'lineparameters')** lets you specify the line style, marker symbol, and line color after the other inputs. The line parameter symbols are enclosed in single quotes and can be given in any order. For example, '-.b' uses a blue dash-dot pattern. Recall from earlier chapters, that **hold on** overlays the next plot over the top of the existing one. Make sure to issue **hold off** later to end overlaying the plots.

>> *syms x*

>> *fplot (sin(x) , '-.b')*

>> *hold on*

>> *fplot (cos(x) , '-r')*

>> *hold off*

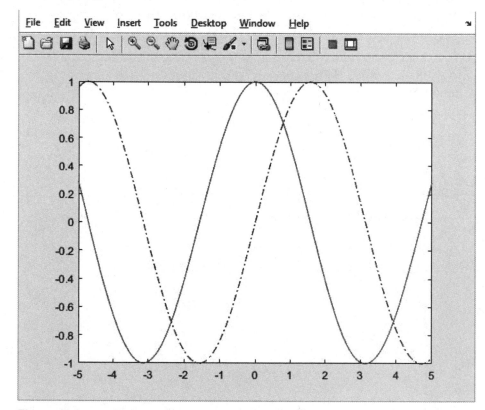

Figure 11.2   **Multiple Functions with fplot**

Clear the Command Window and Workspace.

>> *clear*

>> *clc*

Starting a Live Editor Session

Click: **New button** *from the Home tab*

Click: **Live Script**

Figure 11.3   Home Tab New Button Selections

The Live Editor

Figure 11.4 shows the Live Editor tab. You do not have to use the Live Editor to use the Symbolic Math Toolbox. The *Live Editor* lets you format text and equations, to add description to your MATLAB work. You can also add images and hyperlink to other information. Using the Live Editor you can create complete documentation for your project which includes MATLAB code.

Figure 11.4   Live Editor Tab

Click: **Text button** *(if it is not already selected)*

On your own, enter a couple of blank lines at the beginning of the document by pressing [Enter].

Click: **Equation button**

The Equation tab appears on your screen similar to Figure 11.5.

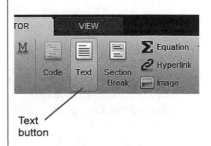

Text button

Figure 11.5   **EquationTab**

Entering an Equation in the Live Editor

We will create a brief paper for the velocity of a falling object as a function of time, with the effect of gravity and air resistance. Part of this brief paper will be descriptive and the remainder will be a MATLAB script that lets us visualize the results. First enter the equation as it would appear in typical mathematical form using the Enter your equation window that appeared when you selected the Equation button. Select the button again if your cursor does not appear in the Enter your equation box.

    Type: **V(**

Figure 11.6   **Equation Parentheses**

Notice that as soon as you type the left parenthesis, a right parenthesis and a highlighted box appear in the equation editor area. This formats as you go to make it easy to type standard mathematical notation as it would appear in a paper or report.

    Type: **t**

As soon as you type the letter t, it becomes italic,

    Press: → *to move the cursor outside the right parenthesis*

    Type: **=**

    Click: Fraction

    On your own click in the upper highlighted box and type mg, then click the lower box and type k.

    Press → twice to move past the fraction.

$$V(t) = \frac{mg}{k}$$

Figure 11.7   **Fraction in the Equation Editor**

Fraction

Tip: *Ensure that you are still in the equation editor. Notice its border is highlighted. If you press the right arrow too many times, you will exit the equation editor. If that happens, click the equation again and reposition the typing cursor as needed.*

Continue to type the equation as shown in Figure 11.9. Before making a subscript or superscript, click the Subsuperscript button and then enter first the value and then the sub- or superscript.

*Click: to expand the Math **Structures** Palette (Figure 11.8). Use the Parentheses button to create a set of parentheses then type inside them.*

Tip: Subscript is a small letter or number below a value, superscript is above. The 3 in x^3 is a superscript.

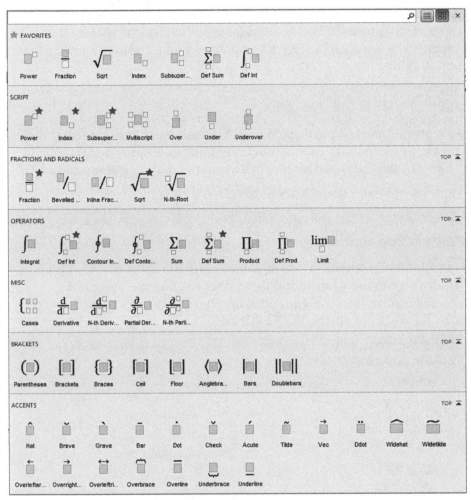

Figure 11.8   Math Structures Palette

Using the structures palette and the symbols palette located to its left, you can enter almost any typical mathematical expression in standard notation format.

*Click: **X** in the upper right of the box to close the palette*

Finish the equation on your own. The superscript for e is a fraction within the superscript. Click in the superscript box and then click Fraction; a new set of boxes will appear inside the other box.

$$V(t) = \frac{mg}{k} + \left(v_0 - \frac{mg}{k} \right) e^{-\frac{kt}{m}}$$

Figure 11.9   Finished Equation

Note: *The constant, (k) in this equation, indicates a hypothesis for linear resistance due to air molecules as compared to the alternative that this resistance is proportional to the square of the speed.*

On your own, save your Live Editor file as Velocity.mlx.

This formula states the velocity as a function of time, $V(t)$, of an object falling through air is a relation between its mass (m), gravitational attraction (g), the constant (k) which is dependant on air density and the shape of the object (essentially the drag constant times the cross sectional area), initial velocity (V_0), and the constant (*e*) which has the natural log equal to one. You may remember the terminal velocity example in Tutorial 1. This is a similar equation for calculating the velocity as a function of time when given an initial velocity, rather than merely falling.

Entering Code in the Live Editor

As a programming language, MATLAB must be able to interpret the syntax for its commands and functions, therefore this is necessarily different than the typed and formatted mathematical expression.

Next we will enter some lines of MATLAB code.

*Press: **Enter** to exit the equation editor or click outside the equation box*

*Click: **Code** button*

A shaded box appears in which you can enter lines of MATLAB code. When you run the Live Script, these lines of code are executed. The descriptive text and equations are there to document and describe only. They are not run as MATLAB code.

On your own, enter the following lines of code into the Live Editor code input box.

```
clear
syms t

m = 1.0
g = 981
C = 0.47
airDensity = 0.001225
k = airDensity * C
v0 = [0 200000 500000 800000]
numv0 = 4
range = [0 10000]

t
for n = 1:numv0
   V(t) = ((m*g)/k ) + ( (v0(n) - (m*g)/k) *  ((exp(1))^(-(k*t)/m)) );
     fplot (@(t) V(t) , range )
     hold on
end
```

Your screen should be similar to Figure 11.10.

$$V(t) = \frac{mg}{k} + \left(v_0 - \frac{mg}{k}\right)e^{-\frac{kt}{m}}$$

```
clear
syms t

m = 1.0
g = 981
C = 0.47
airDensity = 0.001225
k = airDensity * C
v0 = [0 200000 500000 800000]
numv0 = 4
range = [0 10000]

t
for n = 1:numv0
    V(t) = ((m*g)/k ) + ( (v0(n) - (m*g)/k)  *  ((exp(1))^(-(k*t)/m)) );
    fplot (@(t) V(t) , range )
    hold on
end
```

Figure 11.10   Code Input in the Live Editor

We provided four different initial velocities (0, 200000, 500000, 800000) to test in the loop that will run four times. Each pass through the loop will plot the velocity as a function of time. What results do you expect?

Click: **Run All**

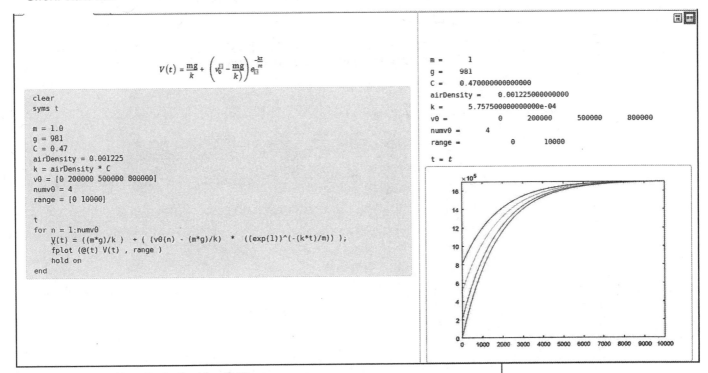

Figure 11.11   Results in the Live Editor

Notice that for each initial velocity provided, as time increases, the velocity approaches the asymptote or terminal velocity.

You can open the plot results in a new window and save them to other formats.

Click: **on the graph image to show the tools (see Figure 11.12)**

? *Try changing to some negative values for v0. Use a similar range of values, for example, -200000, 0, 300000, 600000 so that lines will have some separation on the plot. What is the result? What does a negative value for v0 represent?*

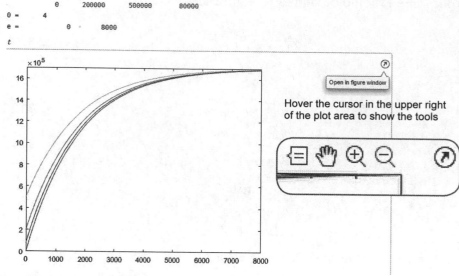

θ = 4   θ   200000   500000   80000
e = 0   8000
t

Figure 11.12   Plot Tools

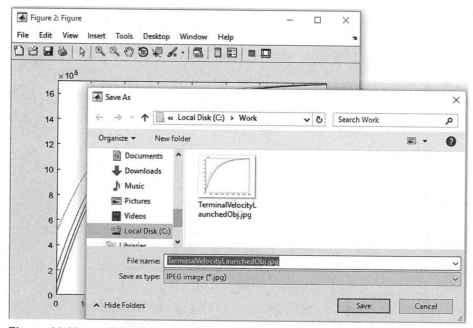

🠑 *Click: Open in figure window button*

The Figure window opens. You can use it to add annotations, save to
other formats and so on. From the Figure window,

Click: File, Save As

*On your own, change the file type to .jpg and name the file
TerminalVelocityLaunchedObj.jpg*

Close the Figure window.

Figure 11.13   Save As Dialog Box

Adding Text in the Live Editor

The text button in the Live Editor lets you enter text. Try it now by adding some description in the blank lines area above the initial equation you entered.

Click: **Text button**

Click: **in a blank line above the equation**

Enter some description, such as:

"Let's examine the velocity with respect to time for an object launched in a direction exactly opposite to that of gravity. The formula for the velocity of an object falling through air as a function of time, given the starting velocity of the object is:"

On your own, click past the equation and ahead of the MATLAB code. Enter descriptive text similar to:

"where m is the mass of the falling object, g is gravitational attraction, k is a constant which is dependant on air density and the shape of the object (essentially the drag constant times the cross sectional area), v0 is the object's initial velocity, and e the constant which has the natural log equal to one."

Your document in the Live Editor should be similar to Figure 11.14.

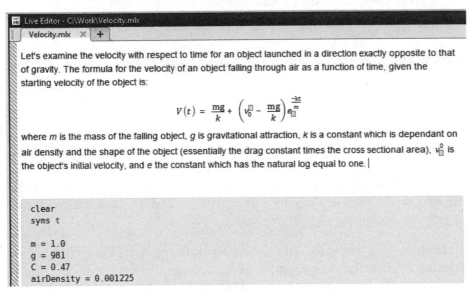

Figure 11.14   Text Added in Live Editor

Embedding Images in the Live Editor

You can also add graphics to your Live Editor documents.

On your own, create a blank line at the end of the last descriptive text you entered. Position the cursor in that line.

Click: **Image button**

A file browser opens on the screen.

On your own, use the browser to select Rocket-1.jpg from your datafiles-matlab folder.

The image is embedded but needs to be resized. On your own, click and drag on the square grips at the corners of the image to resize the figure until it is similar to Figure 11.15.

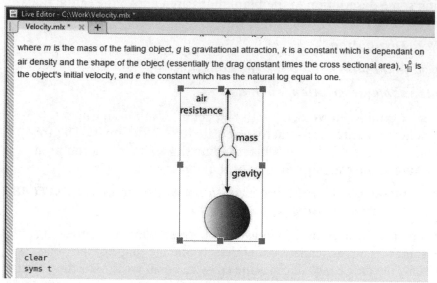

where *m* is the mass of the falling object, *g* is gravitational attraction, *k* is a constant which is dependant on air density and the shape of the object (essentially the drag constant times the cross sectional area), v_0^o is the object's initial velocity, and *e* the constant which has the natural log equal to one.

```
clear
syms t
```

Figure 11.15   Graphic Embedded in Live Editor

On your own, save and close your velocity.mlx file.

Clear the MATLAB Command Window and Workspace.

Heat-Treatment Curve Fitting Revisited

In our heat treatment example, we looked at finding a curve to represent the heating and cooling process. By investigating the rate curves, we may gain insight into which of the curve fitting equations best represents our heat-treatment scenario and to be able to reject parts that were outside the allowed heating and cooling rates.. Now that you are familiar with using **syms** and **fplot**, we are almost ready to compare the first derivatives of the curve-fit equations and use that information to help select among the proper rate curve.

To find the first derivative of the temperature curve we could use basic differential calculus. For example, if the temperature curve is described by the fourth-degree polynomial as follows:

f(x) = p1*x^4 + p2*x^3 + p3*x^2 + p4*x + p5

then the rate curve will be this third-degree polynomial,

f'(x) = 4*p1*x^3 + 3*p2*x^2 + 2*p3*x +p4

MATLAB has a built-in function for finding the derivative, so let's use it. Some of the MATLAB functions for mathematics are:

integral	Numerical integration
integral2	Numerically evaluate double integral
integral3	Numerically evaluate triple integral

quadgk	Numerically evaluate integral, adaptive Gauss-Kronrod quadrature
quad2d	Numerically evaluate double integral, tiled method
cumtrapz	Cumulative trapezoidal numerical integration
trapz	Trapezoidal numerical integration
polyint	Polynomial integration
del2	Discrete Laplacian
diff	Differences and Approximate Derivatives
gradient	Numerical gradient
polyder	Polynomial differentiation

You can read more about them in the MATLAB help files. We will use the diff function to find the derivative.

For our example, the 4th degree polynomial curve fit had the following coefficients: p1 = 0.0008579, p2 = –0.01787, p3 = 0.07824, p4 = 0.4089, p5 = –0.4577.

Use the Command Window to enter the following:

```
>> syms x;

>> p1 = 0.0008579;

>> p2 = –0.01787;

>> p3 = 0.07824;

>> p4 = 0.4089;

>> p5 = –0.4577;

>> f = p1*(x^4) + p2*(x^3) + p3*(x^2) + p4*x + p5

>> g = diff (f, x)

>> fplot (g)
```

You should see the following output for g and a plot in the Figure window similar to Figure 11.16.

```
g =

(494545679401107*x^3)/144115188075855872 –
(5361*x^2)/100000 + (489*x)/3125 + 4089/10000
```

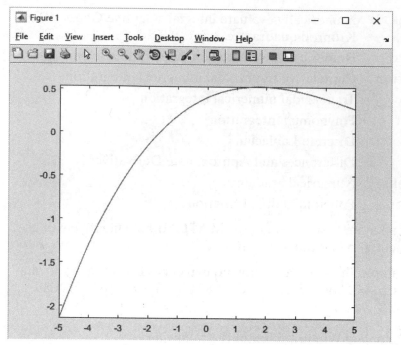

Figure 11.16   Differential Curve for 4th Order Polynomial

But wait, we don't want to type in all those coefficients each time. Plus the default plot starts at -5; that time is before our heat-treating process even started. No sense in projecting that. We'd better set the fplot range.

The script ratecurve.m provides a starting point for generating the rate of change curves.

On your own, load ratecurve.m

```
Editor - C:\datafiles-matlab\ratecurve.m
  ratecurve.m  ✕  +
 1      % Don't forget to load ('heatTreatCoeff.mat') from the datafiles
 2 -    load('heatTreatCoeff.mat')
 3
 4      % Variables
 5 -    syms x
 6 -    rows = size(HeatTreatCoef,1);
 7 -    columns = size(HeatTreatCoef,2);
 8 -    plotRange = [1, 12];
 9 -    rowNum = 2;
10
11      % Process one row of coefficients at a time
12 -    while rowNum <= rows
13          % Use the coefficents from the row of HeatTreatCoef in the equation
14          % f(x) = p1*x^4 + p2*x^3 + p3*x^2 + p4*x + p5   4th order example
15 -        polyOrder = rowNum;
16 -        F = 0;
17 -        j = polyOrder;
18 -        for i = 1:polyOrder + 1
19 -            fprintf("HeatTreatCoef(%d,%d)*(x)^%d + ",polyOrder,i,j);
20 -            F = F + HeatTreatCoef(polyOrder, i)*x^j;
21 -            j = j - 1;
22 -        end
23 %        fprintf("\n");
24 -        g = diff (F, x)
25 -        fplot (g, plotRange)
26 -        hold on
27 -        rowNum = rowNum + 1
28          % Go to the next row and do it all over again
29 -    end
30 -    hold off
```

Figure 11.17   Differential Curve for 4th Order Polynomial

The data file, *heatTreatCoeff.mat*, was created by copying and pasting the coefficients from the curve fit table in the previous chapter. This data file is loaded from the *ratecurve.m* script. To load it, you may need to set the path to the location for your datafiles, or copy *heatTreatCoeff.mat* to your current folder.

On your own, run ratecurve.m, making path adjustments if needed.

The script executes and the Figure window opens similar to Figure 11.18.

Click: **Insert Legend button and drag/resize the legend to a better location.**

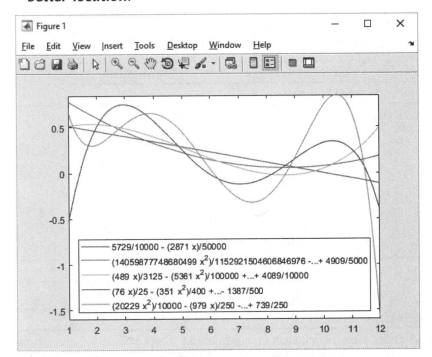

Figure 11.18   Differential Curves for 2nd through 6th Order

What can we determine from examining the rate curves? These curves show the rate of change for the polynomial fitting equations for the heat-treatment data from the previous chapter. Recall that the higher order polynomial equation had the best fit to the data. However, we can see here that the rate of change for higher order polynomial fits fluctuates radically. So keep this in mind when selecting your best fit curve.

In our heat treating example, the process engineer required the part be rejected if the heating rate was found to exceed 600° F per hour or the cooling rate was more than 400° per hour. Examining the curves (and remembering that the heating units on the y-axis are in thousands of degrees) do you think this part should pass inspection or not? Which fit curve are you using for your decision? Why did you select this one?

As you can see, there is some art to selecting the best fitting equations based on deeper knowledge of the processes and relevant science.

Using Symbolic Math with Units

Another great feature of the Symbolic Math Toolbox is the ability to specify units of measurement for the symbolic values in use. Of course you still must be attentive to your input units. Units used with symunit are like symbolic math expressions, similar to the $y = \sin(x)$ function the tutorial started with.

The Symbolic Math Toolbox lets you use either symbolic or numeric arithmetic. Until this tutorial you have been using the default numeric arithmetic. Remember the numeric variables and types of data from Tutorial 2? As it turns out, numeric arithmetic in MATLAB is either *variable precision* or *double precision*. Symbolic arithmetic allows precise calculations and use of symbolic units.

Here's a comparison of the features of symbolic, variable-precision, and double-precision arithmetic. You can find more detail in the MATLAB help under the topic Choose Symbolic or Numeric Arithmetic.

Symbolic	Variable Precision	Double Precision
Example: Find sin(π)		
a = sym(pi) sin(a) a = pi ans = 0	b = vpa(pi) sin(b) b = 3.1415926535897932384626 433832795 ans = -3.2101083013100396 069547145883568e-40	pi sin(pi) ans = 3.1416 ans = 1.2246e-16
Functions Used		
sym	vpa	digits double
Round-Off Errors		
No, exact results	Yes, depends on precision used	Yes, 16 digits of precision

Before specifying units, you must use **symunit** to create a place for them to be stored symbolically. Then you use that storage location with a dot notation (.) to specify the unit (for example, u.mm). You can find the unit types in the help under the topic Unit List. Here are some examples:

Length
.Ao (= Angstroem = angstroem = Angstrom = angstrom) .nm (= nanometer) .My (= micron = micrometer) .mm (= millimeter), .cm (= centimeter) .dm (= decimeter), .m (= meter) .inch (= inch) .ft (= foot), .ft_US (= foot_US) .yd (= yard), mile, nmile, inm (= INM) .AU (= AE), .ly (= lightyear = Lj = lj), .pc (= parsec)
Mass
.ag, .fg, .pg, .ng, .mcg (= mcgram = microgram) .mg (= milligram), .cg, .dg, .g (= gram) .kg (= kilogram) .oz (= ounce) .lb (= pound)

Time
.as, .fs, .ps, .ns (= nsec = nanosec = nanosecond)
.mcsec (= mcsecond = microsec = microsecond)
.ms (= msec = millisec = millisecond), cs, ds
.s (= sec = second), das, hs, ks

Temperature
.K (= kelvin = Kelvin)
.Fahrenheit (= fahrenheit)
.Celsius (= celsius)

Angle
.degree (= angular degree)
.rad (= radian)

If you think there should be a unit for it, there probably is. Use the help system to look them up.

Next, give it a try using our good old terminal_velocity script.

On your own, open terminal_velocity.m in the Editor.

Save the file as terminal_velocitySymUnits.m

Edit the code block to be as follows:

```
% program to calculate terminal velocity
% Vt = terminal velocity
% dia = diameter of sphere (in cm)
% mass = mass of the falling object(65.4710 grams)
% gravity = acceleration due to gravity(approximately
    981.0 cm/s2)
% airDensity = density of the medium (air at 0.001225 g/
    cm3.)
% area = cross-sectional area of the object (in cm^2),
% for a sphere it is pi*radius^2
% C = drag coefficient (sphere is approx 0.47)
u = symunit;
dia=2.50*u.mm
dia = rewrite(dia,u.cm)
area= pi*((dia/2)^2)
mass=65.4710*u.g
gravity=981.0*u.cm/u.s^2
airDensity=0.001225*u.g/u.cm^3
C=0.47
Vt=sqrt((2*mass*gravity)/(airDensity*C*area))
```

*Tip: Remember to take out the semi-colon (;) by the dia = 2.50 *u.mm line and others so you can see the symbolic units in action.*

Run the code you edited by clicking the Run button or type the script name at the Command prompt:

>> *terminal_velocitySymUnits*

The ouput appears in the Command Window as follows:

```
dia =
(5/2)*[mm]
dia =
(1/4)*[cm]
area =
(pi/64)*[cm]^2
mass =
(65471/1000)*[g]
gravity =
981*([cm]/[s]^2)
airDensity =
(49/40000)*([g]/[cm]^3)
C =
   0.470000000000000
Vt =
((7^(1/2)*99952128000^(1/2))/(7*pi^(1/2)))*([cm]/[s])
```

The units appear in square brackets []. Notice the code, dia = rewrite(dia,u.cm), changed the value for dia from its original mm units to cm units. The output Vt = ((7^(1/2)*99952128000^(1/2))/ (7*pi^(1/2)))*([cm]/[s]) is symbolic.

You can convert the value from symbolic to real, but first you must separate the units portion using the built-in **separateUnits** function.

Add the next statements at the end of the code in the Editor:

```
[VtNum,VtUnits] = separateUnits(Vt);

vtNum = double(VtNum)

VtUnits
```

On your own, run the code again.

The following lines appear at the end of the output in the Command Window:

```
vtNum =
   6.741741030699041e+04
VtUnits =
1*([cm]/[s])
```

The symbolic value was converted to a double-precision real number. The units are still a symbolic value. Using symbolic units is a powerful tool you can use to help prevent making errors based on units. Of course you must always track what units the original data uses. There is no foolproof way to prevent unit errors other than paying attention to these details. That said, the symbolic units feature is pretty cool.

You have completed Tutorial 11.

As you have seen throughout this text, MATLAB is an incredibly powerful and versatile tool. This introduction is just a starting point for using MATLAB to create programs and documentation for engineering, science, mathematics, finance, and many other fields. The software has excellent help files to guide you. We hope you use this starting point and continue to explore MATLAB.

Key Terms

Symbolic Math Toolbox

Live Editor

variable precision

double precision

Key Commands

syms

fplot

symunit

rewrite

separateUnits

Exercises

⦿ Exercise 11.1

The following variation of furnace temperature is recorded during the heat-treatment of a part:

Time (Hours after start)	Temperature (Thousands of degrees F)
1	0.102
2	0.458
2.5	.59
3	1.06
4	1.75
5	2.12
6	1.97
7	2.36
8	1.85
10	2.05
11	2.75
12	2.40

The process engineer wants the part to be rejected if the heating rate exceeded 585° F per hour or the cooling rate exceeded 350° per hour. Determine if the part passes or is rejected based on the data observed during its heat-treatment.

⦿ Exercise 11.2

a. Write a script using symbolic units to convert the densities for the materials listed in the table from lb/ft^3 to kg/m^3.

b. Rewrite the script to accept the input of a material name and its density in lb/ft^3 and output its density in kg/m^3.

Material	Density (lb/ft3)	Material	Density (lb/ft3)
ABS	66	Gallium	366
Aluminum	167	Glass, Pyrex	138
Aluminum Bronze	481	Lead	708
Antimony, cast	418	Lithium	33
Asbestos	125 - 175	Magnesium	109
Asphalt, compact	147	Nickel	555
Balsa Wood	8.1	Paper	44 - 72
Bismuth	607	Platinum	1342
Calcium	97	Porcelain	143 - 156
Cement, Portland	94	Silicon	144
Coal, anthracite	87 - 112	Silver	655
Coal, bituminous	75 - 94	Talc	168 - 174
Copper	548	Titanium	281
Cork	14 - 16	Zinc	444
Diamond	188 - 220		

Exercise 11.3
a. Use symbolic variables and fplot to graph the following functions:

a.   sin(x)   Sine of argument in radians

asin(x) Inverse sine in radians

sinh(x) Hyperbolic sine of argument in radians

b.   cosd(x) Cosine of argument in degrees

tand (x) Tangent of argument in degrees

secd (x) Secant of argument in degrees

Exercise 11.4
Read about simplifying symbolic expressions in the MATLAB help. Example:

syms x

simplify(sin(x)^2 + cos(x)^2)

Use symbolic variables and simplify the following expressions.

a.   1/x + x/x-7 + (x+6)/x^2

b.   sqrt (4y/x^2)

Exercise 11.5
Research the physics problem below and use the Live Editor to write a brief (1-2 page) paper explaining it. Use MATLAB code within the Live Editor to provide the solution. Add illustrations of the free-body diagram if you have sufficient graphics tools available.

A 5 g box is on the surface of a frictionless plane included at 30 ° from horizontal. What is the acceleration a of the box down the plane?

INDEX